ARC of CONFLICT

Fraternizing in World War II left scars & mystery in Norwegian community

by

Richard Everett Londgren

© 2015

ATLANTIC
OCEAN

NORWAY

SWEDEN

■Karmøy
●Stavanger

●Kristianstad

Scandinavia
& Baltic Countries

Foreword

Ships, tugboats, barges, log rafts, lighthouses and even sailboats served as the background for our family's many years of living in sight of the Puget Sound of Washington State.

Relating this setting to the Nordic countries— dominated by the ocean and sea—seemed natural to me. But World War II and the bonanza from oil rigs have changed the scene and structure of much of Nordic life.

Thanks to our family's roots in Scandinavia, our visits to Scandinavia, our involvement in the Scandinavian Centers at California Lutheran University (shown above) and Pacific Lutheran University in Tacoma, Washington, and our preparation of material for Scandinavian publications and books, we've added much to our knowledge of our Nordic heritage and history.

All that contributes to this story of the *ARC of CONFLICT*.

Even the name of the central character—Harold Shostrom (Americanized from the Swedish *Sjöström*)—brings together *sea* and *stream.*

Richard Everett Londgren

Introduction

Mind boggling! That's how the German gun emplacement affected my wife Anita and me when we viewed the exterior and then entered the massive World War II structure on the Norwegian island of Karmøy.

That impression deepened for us, because Anita's grandfather, Carl Hillesland, emigrated from that island. But many of Anita's other relatives still live there—a short distance from that German bunker.

Later, we visited the war museum in nearby Haugesund, and a map there left us further astounded: the German invaders had built such fortifications from the northern tip of Norway all along the western European coast to the northern tip of Spain.

The domed structure on Karmøy impressed us in another strange way: we appreciated the design of the domed bunker.

So, out of that experience came this story, starting with the consequences of World War II in Norway and continuing till the country experienced a different kind of shock caused by the discovery of oil in the North Sea.

In a way, this novel serves as an essay about the impact of shocking change. First, in the late 19th Century, nearly one-fourth of the Norwegian population left a desperate life there for a new beginning in America.

Then war.

Now, in an ironic parallel, the Norwegian immigrants succeeded in America, and so eventually have those now prospering in the new promise of oil in Norway that began about 1970.

Chapters

1. Rude Welcome
2. Dinner Dialogue
3. Wartime Review
4. Speech! Speech!
5. Loyalty or Reality
6. Dinner Discussion
7. Picnic with Purpose
8. 4th of July
9. Shift to Sweden
10. Swedish Clan
11. Class Act
12. Recoup & Regroup
13. Haystack Search
14. Tying Loose Ends
15. Welcome Advice

Cast of Characters
Afterword

ARC or CONFLICT

Chapter 1: Rude Welcome

"I will not apologize to Olaf, no matter what you say, Mom!" shouted Jakob, as he threw his backpack and jacket toward the closet. "I've had it with him and his insults, so I gave him a dose of his own language!"

"Well, I just got a call from Olaf's mother," said Mrs. Onstad, "and she was plenty mad about how you cursed at him and threatened to beat up on him. They're our friends, so why are you acting that way?"

"I thought he was too, but then suddenly he starts the Nazi-lover accusations again. And some of the other kids join in. But I didn't curse him. Just called him a bunch of obscenities."

"I can guess," said Mrs. Onstad.

"You can think about it in your room, and we'll talk about it later. First, I want you to meet Karl's friend from the oil platform."

From the living room, Karl shouted for Jakob to join them.

"Sounds like you had a bad day—again, Jakob," grinned Karl. "Well, thanks for trying to defend our honor. I've been there too, so you'll get used to it and get over it.

"Now say hello to Harold Shostrom, my fellow oil tycoon," laughed Karl, with an echoing chuckle from Harold.

"Interesting to meet another member of your family," Harold declared. "And I heard what you said, and I've been there too."

"Sorry that I exploded like that," said Jakob, "and I'm glad to meet you. American, aren't you?"

"Even after I've been working with your brother, you don't think I sound like I'm Norwegian?

"On the other hand, you could have guessed Swedish. Maybe my name anyway, if not the sound," laughed Harold, as Jakob and Karl joined in the ethnic humor.

"Well," said Jakob, "I'd better head for my room…or get in even more trouble. I'm curious about the problem you mentioned. Maybe later we can talk over our troubles."

"Helps to talk about it," said Harold. "Some of that frustration I felt as a kid still upsets me, and I want to go back and kick some butt. And now I'm big and tough enough to do it," he laughed.

As Jakob left, Mrs. Onstad joined Karl and Harold. "First," she said, "my name is Olga…sometimes *ogre*…to my family," she chuckled. "And you can stay for dinner, I hope. Maybe it'll be a calm meal, though Jakob, Karl and their dad do become a bit agitated at times.

"But before I go back to work in the kitchen, maybe you and I, Karl, should fill Harold in on what Jakob referred to—that Nazi insult."

"It does haunt our family," admitted Karl. "I'll give you the short version now, and you may or may not want to learn more later.

"Our cousin Hanna, daughter of our dad's sister Gerta, had a German officer as a father," explained Karl. "Gerta insisted that he was an outstanding German, not a Nazi. Said he was an architect and engineer sent to direct the coastal gun emplacements here and all up and down the Atlantic coast. According to her, he was a kind and considerate companion to her in during the early days of German occupation, before the later brutality of the Gestapo intervention. He even taught design to some of her best students.

"Alas, he was handsome as well. Not greatly surprising…that, unfortunately, led to romance."

"And, also unfortunately, she became pregnant," added Olga. "Maybe you know—Harold—that in Norway, fraternizing with Norwegians even from other towns is resented. So fraternizing with the enemy is strictly forbidden."

"I get the picture…and the message," interjected Harold. "Surprisingly similar in our changing American mix, as people from neighboring towns and certainly from differing religions and ethnic backgrounds still tend to stay separate. And continuing problems about color, as I'm sure the whole world knows.

"Strange, many prisoners of war from Germany and Italy did stay and mix in, even worked on farms when help was needed. Some came back to the United States to marry a woman from the community. Maybe not with total local approval though.

"Later, thousands of others called 'displaced persons' were rescued and brought to America, largely to evade Soviet clutches. Unlike earlier immigrants, such as my great-grandparents who could hardly wait to leave Sweden to get a better life in America, many of these later immigrants still longed for their homelands, so they hang together in their homeland clusters. And experience some resentment in America."

"We learned here that many Norwegian 'freedom fighters' who were trained and based in America and Britain started a second family there," said Olga, "silently forgiven, while their Norwegian families were left to fend for themselves. Unfortunately, that's an embarrassing part of our heritage too.

"On the other hand, punishment of fraternizing women here turned vicious," explained Olga. "Other women berated and bullied Gerta, even shaved the hair off her head, and labeled her a traitor.

"That extended beyond the war, because Gerta's daughter Hanna has suffered similar treatment. So they both fled to Sweden…with help from Col. Schorn and our family."

"Now, Harold, you probably can see the picture of how the Nazi accusation doesn't go away," said Karl. "But at least Greta and Hanna escaped to a better life, and Hanna teaches in a college in Sweden. Greta retired from teaching and bought a small farm that she and Hanna share, though Hanna also has an apartment in Kristianstad. I see them once in a while, usually there, because they don't relish coming back here.

"Sweden accepted Greta's teaching credentials—which she augmented to meet Swedish requirements. And Sweden accepted her reasons for leaving Norway.

"And, of course, Hanna matriculated in Sweden, all the way to her advanced degrees—to become a college professor."

"Sounds intriguing. I'd like to get their personal side of their story someday," said Harold.

"Maybe during my next work break, if you can get away to go along to Sweden," said Karl.

"Sounds good. Perhaps we can arrange to visit some of my Swedish relatives, too," said Harold. "I met some of them when they visited in Seattle, and I really grew fond of them.

"Time to build on that relationship," he added, "and maybe connect your relatives with mine. After all, we Scandinavians, especially in the Ballard area of Seattle, learned to live together peaceably. Even to fraternize," he laughed.

As his mind drifted back to Ballard, Harold thought about his life—a shaky start as a premature baby, growing into a studious runt, next progressing into a studious athlete. Outstanding pals, destined for success, kept him on his toes. Girlfriends were intellectual and independent…educated and focused, like his former fiancé.

"Harold…Harold…he heard, as he felt a nudge against his arm.

"You went into a bit of a trance," laughed Karl. "Back to America, I imagine."

"Yeah, I guess seeing Jakob so upset shifted me back to that age," explained Harold. "Never had bad dreams like those of Jakob, though.

"Thanks for bringing me back to the present."

Chapter 2: Dinner Discussion

At the dinner table, Harold marveled at the array of food, exclaiming "sure beats my home-cooking and the company cafeteria. As a scientist, I pronounce this nourishing as well as taste-tempting."

"Well, Dad, I think she's outdone herself!" declared Karl, as Olga smiled.

"Yeah, and most of it local from here in Stavanger," pronounced Haaken. "The salmon, of course. Even the vegetables, despite our short growing season. Dairy products, thanks to our goats. And our strawberries."

"Yum!" added Jakob. "Especially glad to have you here, Harold. Brings out the best in Mom's cooking!"

"Well, I'm mighty glad to be here for this feast and for the enjoyment of being with your family," Harold responded. "By the way, Jakob, Karl told about the war-time incident in your family, and now I understand the anger you feel about of being called a Nazi.

Stavanger International Church

"I've learned about some of that past, as well as about the past of long ago, in the discussions after Christian worship services set up here for us foreign oil-workers," said Harold, to indicate his growing knowledge of Norway.

"I particularly like my namesake, Harald the Fair-Haired," reported Harold, with a smile. "Your King Harald managed to defeat some of the Norwegian fear of fraternizing by uniting all of your remote villages and cities into a nation. On your island of Karmøy, I learned that he made a good living by demanding a fee from ships going through that narrow strait connecting with Haugesund."

"Pretty shrewd, huh!" declared Jakob. "If he had been around now at the time of our oil discovery, he certainly would have understood its value."

"Good thing we have our Fred Olsen, right here, to recognize such opportunity," announced Haaken. "One of the first to see our oil potential. He is a bit like King Harald in timeliness," added Haaken, with a slight grin.

"Oh now, I sense a pun there," laughed Karl.

"Why is that?" asked a puzzled Harold.

"Fred Olsen has built Timex watches into a world business," explained Karl. "He doesn't waste a second in his businesses....

"But how about your career, Harold. From what I've gathered, you've slowly and steadily ticked your way toward success."

"Oh, I've done whatever comes my way," Harold answered, ignoring the puns. "Delivering papers as a kid, about your age, Jakob. Fourteen I'd say."

"And I'd say you just won a prize," laughed Jakob.

"Then I worked in the printing department for a small local newspaper in Seattle. Took photos and wrote stories as well."

"You worked for a printer!" exclaimed Jakob, as he interrupted Harold.

"Sounds as though you're interested in printing," said a surprised Harold. "Good for you. Great way to learn about writing and design."

"I'm interested in another way—so is our family, at least Greta and Hanna are," explained Jakob. "They have a mysterious poster from the war that her German officer wrote on to serve as a will making Greta heir to some sort of treasure. Signed by him and some other Germans. They're gone, of course. And also witnessed by a Norwegian—well a 'quisling' who was later executed."

"We have no knowledge of a specific treasure," Karl stated, "and this might be just some Nazi practical joke. But it causes more harassment for Jakob because he talks a lot about it at school."

"Well, I believe it!" shouted Jakob. "And I intend to solve this mystery—even if it takes the rest of my life!"

 "My sister assured me that she thinks it's real," said Haaken, "but I don't know if such a will would stand up in court—even if we could track down some hidden valuable. And would it be confiscated if we did!

"I want all of you to know that Greta truly believed that her German officer—his name was Col. Ludwig Schorn—was a noble person. From Mecklenburg, where the Germans weren't dedicated to Hitler—but loyal to Germany."

"He was an architect assigned to building the coastal fortifications," continued Haaken.

"But why did the Germans think the coast of Norway was important enough to set up artillery here?" wondered Harold.

"Hope you're patient enough to listen to my long explanation," answered Haaken. "And, I admit, I heard a lot of the reasons later as I began to understand the situation.

With cunning, bluster and bullying, Adolf Hitler (left) directed German aggression, aided by Joseph Goebbels (right) as his propagandist.

Heinrich Himmler (inset) led the feared Nazi Gestapo's brutality in treatment of Norwegians near the end of World War II.

"Well, much connects to your Swedish roots," stated Haaken, with a wry smile.

"Germany counted on iron ore from neutral Sweden to produce the vital steel for their military might. But the most dependable transportation called for rail shipments from Sweden to our port of Narvik. Because of the Gulf Stream, that port was open all year, while the Baltic Sea froze down the Swedish ports."

"I can see that shipping along the Norwegian coast put the ships at risk of attacks by the British," surmised Harold.

Though the aristocratic leaders Winston Churchill (left) of Great Britain and President Franklin Roosevelt of the United States were known for their political and communication skills, they avoided any personal meetings with the despised Adolf Hitler and his team—that might have curbed Nazi aggression.

"Probably," pondered Haaken, "but, according to what I heard later, the British leaders at that time put more energy into political infighting than to actual military fighting.

"No pussyfooting by Hitler, who badgered and bullied…and concocted situations to serve his plan to run Germany."

"I guess that preparation occurred, in contrast to the British lack of organization and leadership," surmised Harold.

"And I imagine the German technology prevailed despite war setbacks," added Harold. "So the Germans had been efficiently building and organizing their military muscle while the rest of Europe slept. Strange, considering how the first World War practically wiped the Germans out."

"I understand," explained Haaken, "that the peace treaty excesses of punishment to Germany also fired them up, using propaganda about how Germany had been mistreated by Britain and France. The Nazis even invoked the music of Richard Wagner as inspiration."

"And I learned that the United States wasn't much help in stabilizing Europe, despite President Wilson's good intentions about the new League of Nations," admitted Harold. "And after that, our country wanted no part in the problems of Europe."

"Alas, I'm afraid that was true," said a saddened Haaken.

"Thanks, Dad," said Karl, as Haaken wound down with his speculation. "I learned a lot that was too much of a bother to think about before."

"Ditto from me," echoed Jakob.

"Thanks for listening," Haaken commented softly.

"Well, back to Col. Schorn," he brightened.

"We learned that in 1944," explained Haaken, "that Hitler had demanded reinforcement of what he called the Atlantic Wall, which included a series of fortifications like the one on Karmøy. That 'wall' went from the northern tip of Norway to the northern tip of Spain.

Field Marshall Rommel

"So Col. Schorn was sent to France to support the work of German Field Marshall Erwin Rommel in beefing up the Atlantic Wall in anticipation of an Allied invasion."

"Let me give you a break in your explanation," interjected Karl. "And correct me if I'm wrong, but we learned later that Col. Schorn died during that subsequent Allied invasion. So we obviously missed that vital personal link to the poster he inscribed for Greta.

"Regarding that poster, we also learned that the poster came about because he had handled an earlier additional duty," explained Karl. "As a designer and from a family involved in printing, he had been assigned the responsibility for distributing propaganda posters while he was stationed in Norway. So he had posters printed, like the signed copy Greta has. As I recall, it simply says 'NORDICS, join our German battle against Communism!' A simple appeal at first. Of course, the threats to and assassinations of Norwegians came later."

When Harold mulled over the prospect of a mysterious treasure as he joined the family in coffee and dessert, he commiserated with Jakob and the rest of them about the underlying stress they all must feel.

"I've seen the movies about the Norwegian resistance during the war, which was especially important in delaying German development of the Atomic Bomb," Harold stated. "And I've read a lot about the Allied invasion in Normandy and the War Trials after the war."

"Yes, our people experienced terrible hardship, hunger and brutality during the war," said Olga. "So we thank America in particular for liberating us and restoring our life as we knew it before."

"Now," offered Harold, "I'll do what I can as a listener for comments and clues that might be helpful with your current challenge. As an outsider, I'll be alert for scuttlebutt that might seem unimportant locally but could catch my own attention in a meaningful way."

"First," said Jakob, with a grin, "what is 'schuttlebutt'? More Western slang?"

"Navy slang, from my days with the Navy Seals," said Harold. "Means minor news being shared, like gossip."

"Wow! The Navy Seals!" exclaimed Jakob. "I've heard of them. Famous for taking on the worst of challenges!

"Maybe you could help train me to be ready when the bullies attack me."

"Sure. Glad to share some of the tricks of the trade," responded Harold. "Keep me sharp and in shape, too."

"Wait a minute!" interrupted Olga. "I hope you aren't planning to break Olaf's arm."

"No, I'll just help Jakob dis-arm him, so to speak," chuckled Harold.

"Just remember, we're trying to heal, not hurt," said Olga.

"Well," interjected Haaken, "sometimes a show of strength can promote peace."

"Good to hear you laughing again, Jakob," said Harold. "Now I think it's time for me to skedaddle."

With that, Jakob, in mock exasperation, bumped his forehead with the palm of his hand.

"Till we meet again," said Harold, as he shook hands all around.

"Well, thanks again," said Harold. "Time for me to amble along, so *adios amigos*, Jakob. *Hasta la vista.*"

"I know those words from your Western movies," exclaimed Jakob. "See you soon, friend!"

After Harold left, Olga expressed her appreciations for Karl's friend and colleague. "A pleasant and smart person, not a 'rowdy' as some of oil workers are called."

"Maybe we could go to a Western movie with him sometime," said Jakob. "He could explain some of the cowboy words."

"I vote for that," said Haaken. "He might welcome some company as he tries to fit in here. Not easy mixing in with a bunch of Norwegians…even for a friendly and outgoing American."

"Several other Americans like him are here, so that helps," explained Karl. "Also, he gets together with Anna Hansson when she gets time off from her work as a nurse on the oil platform."

"Good!" said Olga. "I like her. Sassy but smart."

"And very attractive," added Haaken.

"Kari and I are trying to arrange a double-date, as they call it in America," added Karl. "If our schedules ever work out.

"He's done some skiing in the mountains around Seattle, so he's eager to try that. Downhill or cross-country. I'd have to get in shape again," Karl chuckled. "Another kind of double-date."

"And I want to introduce him to my teacher, Miss Stensgard," announced Jakob. "She's young and good-looking and smart."

"Good thinking, Jakob," commended Karl. "And she's a friend of Kari."

"Well, I hope he doesn't have a girlfriend in America, to scuttle your matchmaking," smiled Olga.

"He did have," answered Karl. "He told me his tale of woe. Seems that moving to Norway didn't appeal to his girlfriend from the university. His former girlfriend."

"Maybe he can find true love here," smiled Olga.

"Sounds like you want to be a matchmaker too, Olga," laughed Haaken.

"In the interest of international relations," she laughed.

Chapter 3: Wartime Review

The next day at his office, Harold turned to his Norwegian assistant, who earlier had helped him get the feel for life in Stavanger when he moved there.

Matronly, educated and resourceful, she said she had literally grown up with some of the family, when Harold asked about the Onstads.

So Marit Dahl explained that the war years turned the country upside-down. First, the shock of the unexpected invasion by the Germans, then an easing off as occupation settled in, followed by desperation as the Gestapo took over to brutally counter the rising resistance by Norwegians and prepare for a possible Allied attack.

"The Nazis conscripted able-bodied men, even shipped them out to work at distant factories or to concentration camps," explained Marit.

"Guess that's the reason I remained single," she explained, with a wry smile. "Most of the young Norwegian men were gone—and that's why I think some women found the young Germans appealing. After all, they had been accustomed to a steady stream of German tourists before the war.

"But Greta Onstad was special," said Marit as she thought back. "She was educated, smart, attractive, sophisticated—an outstanding person, except to women who were jealous of her.

"No wonder Col. Schorn found her appealing and lavished attention on her," explained Marit.

"Then I imagine he saw the writing on the wall when Greta became pregnant, so he evidently had arranged for his family in northern Germany to find a place for her and the coming baby in neutral Sweden.

"After all," explained Marit, "Sweden had strong connections with Germany…in business, even in their related royalty. And the distance from northern Germany to southern Sweden was not great.

"So Col. Schorn, not long before he was transferred to his coastal assignment in France, applied his influence to pave the way for Greta and their baby to travel to Sweden.

"But not before she—they—suffered harsh treatment in Stavanger," added Marit. "And that too is understandable, given the fear, anger and suffering among Norwegians who had lost their independence…and even their sources of food."

"Thanks, Marit," said Harold quietly. "You provided a helpful, sobering slice of history. It was before my time, but at least in Seattle, we did hear about the ordeals of Norway during the war."

Chapter 4: Speech! Speech!

Jakob wasted little time the next day at school in reporting about his cowboy friend from the Wild West in America. He breezed past Olaf without paying him any attention, even when Olaf showed his "trademark" clenched fist.

The other kids listened eagerly when Jakob told about the Navy Seal who worked as an "explorer" in the oil business.

When the teacher, Miss Stensgard, got wind of the enthusiasm of Jakob and his "audience," she asked one of students at the edge of the group what was in the air that generated so much interest.

"Jakob has met the American friend of his brother, so now Jakob has all kinds of stories about the Navy Seals and the Western movies and the Pacific Ocean," said the young girl. "You know how excited he gets about something unusual. He's even done some research in geography. It's fun to hear him talk. I wouldn't mind meeting a Navy Seal myself."

"Well, maybe we should have Jakob do a 'show and tell' this afternoon, like students do in America. I wouldn't mind hearing about a Navy Seal myself," she chuckled. "Would you ask Jakob to come over and I'll ask him to tell about the Navy Seal after lunch."

The girl did, and Jakob welcomed the opportunity.

When talking with Miss Stensgard, he started sharing some ideas.

"I did a little research after I met him. Guess we could start with a map so I can point out Harold's home in America," said Jakob. "I think Seattle, where he's from, is quite a bit south of Stavanger. More like Paris. But similar weather to ours, I think.

By the way, his name is Harold Shostrom. In Swedish that means sea and stream, I found out."

"Seems fitting," smiled the teacher.

Thus, in a few minutes, Jakob relished his place at the front of the class, as he pointed out on a large world map the locations of Seattle and other cities of the West as they relate to Norway.

"San Diego," he explained, "is where the Navy Seals train."

"I'd like to know more about that!" shouted Olaf, as other heads nodded in agreement.

John Wayne

"His office is near here," Jakob said. "I'll ask if he would give a 'show and tell' also. He kinda looks like a star in one of the Western movies. Tall, like John Wayne. Talks like a cowboy sometimes, too. You know—rustle up some chow…so we can corral some rustlers."

"I'd like to know more about Hollywood, too," giggled one of the girls.

"I assume he's met some stars," replied Jakob. "Maybe even Marilyn Monroe. I think her father was from Norway. Maybe Stavanger."

"With that dramatic announcement, I think we should clap to thank Jakob," said the teacher. "I believe they do that in America."

Meanwhile, a couple of weeks later, Harold came back to the Onstads for another visit.

He had hardly arrived when Jakob made his pitch for Harold to talk to Jakob's class about his life and career. Jakob explained that he had recently given a "show and tell" about Harold and the Navy Seals.

"Miss Stensgard, our teacher, agreed with me that the class would like to learn more—directly from you.

"By the way, she's young and very pretty, so I thought you might like to meet her. No ring on her finger either."

"He makes quite a sales pitch, I'd say," chuckled Haaken. "I wouldn't mind hearing what you have to say, myself. Maybe you'd even be willing to offer your 'show and tell,' as Jakob calls it, to the community in an evening event. Could be at the school."

"Hold your horses," laughed Harold. "Seems to me that my spokesman Jakob has pretty much told my story already."

Then Karl's girlfriend, who was also visiting the family, spoke up. "I'm Kari Larsson, friend of Karl and his family and a friend of Jakob's teacher. She mentioned enjoying Jakob's report and liked his suggestion that her class hear directly from you. And expanding it to include others sounds desirable. Good for international relations, too," she concluded, with a hint of a smile.

"Looks like they're ganging up on you, Harold," laughed Olga. "But I want to hear more of your story too. And we're always eager to learn more about America."

A few days later, Jakob dropped by Harold's office to tell him of a meeting he had arranged for Harold to talk with Miss Stensgard.

"She likes the idea of a presentation open to the public as well as for the students," announced Jakob. "She suggested that you stop by at the end of a school day when it's convenient to discuss your project, if that fits your schedule. And maybe a little earlier so my class gets to meet you first," added Jakob.

"Okay," said Harold, as he checked his desk calendar. "Would Wednesday work for her…and you?"

"I think it would," said a confident Jakob. "But I'll make sure and let you know."

On Wednesday afternoon, Harold hiked the four blocks to the school, where Jakob waited eagerly.

"Good!" he said. "I'm glad you could come early for our 'show and tell' with the class before school's out."

"Our 'show and tell'?" chuckled Harold. "I thought you were just setting me up to face a bunch of your classmates."

"Well, I thought that might not be fair," said Jakob, "so I can help monitor the questions. Maybe explain them too, if needed."

"Sounds like good teamwork," grinned Harold. "Go ahead, lead me to slaughter."

When Harold saw that Jakob was puzzled, he explained: "A saying, like leading a lamb to slaughter."

"More Western talk?" questioned Jakob.

"Nope, Bible talk," smiled Harold. "From the Book of Isaiah."

"See how much I'm learning from you," grinned Jakob.

"Okay, here we go…it shouldn't be too baaaad," Jakob laughed.

"Not funny," answered Harold, with a smile. "I'm the one to be slaughtered!"

In the hallway, Miss Stensgard, a bit startled by the big American, introduced herself and thanked Harold for coming.

"The students are excited, eager to see and hear you. Afterward, maybe you and I can go to our teachers' lounge for coffee…and discuss the possibility of a presentation by you at a community gathering."

"Okay, to the slaughter, Jakob," announced Harold.

Then Jakob explained the "sheep led to slaughter. Not the Wild West. The Bible."

Miss Stensgard shook her head slowly from side to side, wondering what kind of 'show and tell' to expect.

The shock of the students exceeded that of Miss Stensgard's, as Harold, attired in his Levi jeans and jacket and cowboy boots, put on his 10-gallon hat as he entered the classroom.

"Wow!" exclaimed Olaf. "You look about 10-feet tall…so how tall are you?"

"Only 6-feet 4-inches," answered Harold. "Same as John Wayne, when I stood next to him when he was acting in a Western. I suppose you've heard of him."

"Yes! Yes!" said one girl, "and you even look like him."

"Played football like him, too," Harold declared.

"Where?" asked a boy.

"Me, for the Washington Huskies," said Harold.

"Is that because the

31

players are husky like you?" asked another girl, feeling confident.

"Good guess," laughed Harold, "but Huskies are beautiful sled dogs, like those in stories by Jack London."

"Oh yeah! We've read his exciting stories about White Fang, the wolf-dog of Alaska," said another boy.

Then Jakob pointed on the map that Stavanger is as far north as parts of Alaska.

"Have you been there?" came another question.

"Yup!" answered Harold. "I worked on a commercial fishing boat in the summer while I was a student at the University of Washington. That made me appreciate going back to the books.

"The Pacific Ocean there is as bad as your North Sea, where I've had to dive down to check out the structure of an oil

platform. Many others died doing that, just as a few fishermen did in the storms off the coast of Alaska."

"Is that how you became a Navy Seal?" asked another student.

"No, after I graduated from the University of Washington, I had to serve in the military, so I chose the Navy Seals. And that training was even tougher than practicing for football. Most new Seals couldn't take it and dropped out after the first week, but I endured and went on to installing underwater explosives during the last war. Many others died doing that. That certainly made me appreciate being in the classroom and studying from books."

"Time's up!" announced Miss Stensgard. "If you have more questions, save them for later, when we hope Dr. Shostrom will talk again in a meeting where your parents and others from the community will be invited."

"Will we be able to attend too?" asked Olaf.

"Well, you are part of the community," smiled Miss Stensgard.

"Maybe, in fact, Dr. Shostrom may want some of you students to help in his presentation. Especially you, Jakob, because you got the ball rolling by enlisting him as a school presenter."

ARC or CONFLICT

Chapter 5: Loyalty or Reality

In the school lounge, several of the teachers lingered.

"I guess they are hoping to meet the notorious American visitor," whispered Miss Stensgard. So she introduced him to the others, and she gave their names…which he immediately forgot.

"Let's get a cup of coffee if you'd like," she said. Then she smiled: "My name is Sofie. Okay to use that now that we're out of the classroom.

"Many thanks for sharing about yourself and your life. I could tell that the students took it in like a sponge. And so did I.

"If you're willing to make a presentation for the community, I can help you prepare, now that I've got the gist of your background."

"Thanks, and that could be interesting…and perhaps serve some good in the community, too," said Harold.

"But if you can spare some time now, I'd like to know more about the festering fraternization issue still affecting the Onstad family."

"First," started Sofie, "I want you to know how much I regret that my grandmother and mother were involved in the cruelty toward Greta. I understand that Greta was guilty of fraternizing with the enemy, but I've heard many positive reports about her Col. Schorn.

"Thanks," responded Harold. "Don't know why I wonder so much about folks here, because we certainly have a mountain of crass behavior like that in our country.

"At the same time we proclaim our Judeo-Christian heritage, we perpetuate senseless conflict between families, communities, and many other groups."

"I suppose you don't know how much unification frustration Norway has experienced for centuries," Sofie offered. "At one time our country lived with Germany's oppressive Hanseatic League, then the demands of the Kalmar Union of the three Scandinavian countries, next Danish domination, then almost a hundred years under the thumb of Sweden, followed by the Nazi invasion.

"Along the way, King Harald knocked heads together to form a nation out of independent Norwegian villages and cities.

"So we do tend to be a bit skittish about outsiders. As a result, different parts of our country hang on to their own customs, and defend them from encroachment by others. Such as regional designs of folk dress called bunads.

"Even different areas of our country claim unique decorations known as rosemaling. Protected like trademarks, such as your Nike swoosh, which looks a bit like a brush stroke of rosemaling.

"I know, because in my family's home area up north near Trondheim, my relatives and their friends police and defend their identity."

"I imagine your Viking era forced quite a lot of change," surmised Harold, "with the importation of ideas, materials and even captives, who might have helped offset the risk of inbreeding in your remote communities."

"With families encouraging marriage of cousins, because those were considered known qualities, inbreeding would have resulted—no doubt with mixed results eventually," said Sofie.

"But some of our cultural leaders saw the need for change," Sofie added. "Henrik Ibsen wrote in critical as well as traditional ways about our trolls, and he criticized our stubbornness in his play about the doctor trying to stop tourism business when a sickness struck the town.

"Amazing coincidence that Edvard Grieg lifted folk music to a higher level, in which he also raised concerns about our traditional trolls. At the same time, Edvard Munch shocked our people and the art world with his mystic paintings," concluded Sofie.

"And now your people have to put up with the mix of outsiders like me involved in the oil business," smiled Harold.

"We are accepting the reality that there is nothing as sure as change, even here in Norway," admitted Sofie.

"How did you yourself manage to adapt to change in face of the deep-set traditions?" Harold asked. "If I'm not getting too personal," he added.

"Not personal," replied Sofie, with a smile of reflection about her past. "Just historical, mainly I adjusted when I attended college in Trondheim. I saw quite a range of students, largely because Trondheim was changing in step with the petroleum industry.

"My cadet teaching at an outlying school also helped make me aware of growing differences in Norway. Not only aware," Sofie added, "but appreciative of the differences."

"Well, the immigrants to America tended to settle with their kith and kin, especially in small towns and rural areas," Harold

admitted. "Norwegians in particular avoided Swedes…and certainly avoided Roman Catholics.

"Meanwhile, I appreciate your input," smiled Harold. "I do believe it will serve me well in my life here in Norway. I have to understand and adapt to change also.

"Now, when it comes to the challenges in the Onstad family, I hope you can help them and me deal with their festering past. Karl has invited me to go with him to Sweden to meet Greta and her daughter, Hanna, so any further preparation before then from you would be helpful.

"And before I meet them, I would like to know more about the improvised document designating Greta as the recipient of something mysterious indicated as a valuable item. Jakob, for one, is convinced it relates to a treasure."

"His one-track mind on that situation does cause stress," explained Sofie, "because several of the kids and others in the community scoff at his belief and make fun of him. Most people, even Norwegian officials, conclude that the Nazis wouldn't allow any valuable jewel, art or artifact to escape their intense screening. So almost all except the Onstads dismiss that wild dream about a hidden fortune. And even most in their family consider it a fairy tale.

"Naturally, many families here know about the Greta's fraternization with a German officer and about their daughter Hanna. Some of the students goad Jakob with that information, so that is an embarrassment that alienates him from many classmates. But the story of the treasure helps him deflect that kind of verbal abuse."

After Harold thought about the value of a presentation about his role in Norway and the preparation for it, he concluded that the community could benefit from his story and the rapid change in local society being caused by the petroleum industry.

So he decided to accept Sofie's offer of help in preparing an appropriate presentation. Besides, he admitted to himself, he liked her and looked forward to getting better acquainted.

He felt sure he would get significant input from colleagues at his office, as well as benefiting from his company's capabilities for creating and supporting a presentation.

And Karl and his friend Kari would certainly provide helpful local insight as well as encouragement and assistance based on their background, he in engineering and she in pre-law studies at the University of Bergen.

Stavanger

And they are known and respected in Stavanger, which, Harold realized, would provide credibility and connections in the community and help draw people to a presentation.

In checking at the office, Harold not only found a 35mm projector, but a colleague skilled at using it for presentations. When his colleague heard the purpose of the proposed gathering, he volunteered to prepare basic charts, title slides and cartoon characters to help share the message in a humorous way.

"Hey," exclaimed Harold as he thanked his colleague, "we've got Walt Disney right here in our office. I sure appreciate your understanding of our purpose. And that you see the need to add some touches of entertainment to help the message sink in.

"Now that I know I have you on tap as such a skilled helping hand," added Harold, "I'll be back in touch to work with you, just as soon as our team zeroes in on the points we want to make."

Chapter 6: Dinner Decisions

Sofie didn't expect the idea of a "The State of Statoil" presentation to catch on so fast in the Stavanger, so she felt guilty about putting the burden on Harold.

"I at least want to help by hosting a dinner involving you," she said to Harold, "with Karl and Kari as well, so we can share ideas with you."

"And I readily accept," answered Harold. "Should be an enjoyable way to launch the planning for the presentation. And if you are as skilled as a chef as you are as a teacher, I look forward to savoring an enjoyable meal as well as an interesting and informative discussion."

"Don't expect an old and traditional meal," Sofie stated. "More and more, we have new traditional meals as Stavanger becomes increasingly international. So I will do what I can to combine past and present," she smiled.

A few days later, when Harold arrived at Sofie's apartment, he didn't know quite how to greet her. He thought a handshake would be too formal, a kiss too personal. So he hugged her. And she reacted with a hug and a kiss on his cheek.

"I like that response as a growing American custom," he smiled.

"Here too," Sofie replied warmly, "despite our traditional preference for privacy. Good to feel such a warm touch of friendship."

"I agree," said Kari, as she came from the kitchen and hugged Harold. "We're cooking up a storm…though you might prefer a term not so much connected with your maritime life."

 When Karl arrived and the four sat at the dining table, Harold expressed his appreciation for Sofie's apartment.

"You've done wonders in making your place so attractive and seem so spacious. You'll have to take me later on a tour of your photo 'gallery' and see if I can guess who they might be."

"Old memories from home and new memories from here, mostly from school and church," Sofie explained briefly, as she helped serve the meal of salmon, boiled potatoes, coleslaw and rye rolls.

"I hope," she said with apology, "you don't mind having salmon that came from ocean farming."

"Not at all," replied Harold. "I know that's a major business here, and it's increasing in America, especially in my Pacific Northwest. You probably have heard of the aviator Charles Lindbergh," continued Harold. After he noted the affirmative nods, he explained: "His marine biologist son Jon leads in the Pacific Northwest as a pioneer in raising fish as a form of agriculture. In the ocean, rather than over the ocean like his father. Maybe even here in Norway."

"I say 'cheers' to that kind of change and to our visit and meal here tonight," said Karl as he raised his glass, as the others responded with a gentle "clink" of glasses. "For our growing friendship, and for the growing communication we hope for in our community."

During the savoring of a lingonberry dessert and coffee, Sofie mentioned that she had invited them all as another step in building a connection between the petroleum business and the community. "I've also informed Harold about the personal complications associated with the German invasion and how that has involved and affected our families."

"I do have to admit," said Harold, "that I'm as puzzled as a lot of others seem to be about the German officer and your aunt, Karl. I appreciate your invitation to visit your aunt and cousin in Sweden. That might provide me some answers to the puzzle of the will written on a poster. And some insight into the understandable bitterness of your aunt and cousin."

"Maybe by having an outsider such as you ask questions," responded Karl, "our family might gain from fresh insight about what happened during the war."

"Before we go our separate ways," Sofie laughed, "let me remind you of why I called this meeting. We need a place to hold a large meeting for the community, with equipment such as projector and microphone.

"Well, I think I can get the school auditorium, and Harold, you indicated you can borrow a projector and microphone. Then we still need to invite the community."

"As you know, I work for a large, old printing company," said Kari, "so, with your help, Sofie, I can arrange for promotional flyers."

"And we could distribute them through the schools," said Sofie.

"Stop the presses!" interjected Harold, with a chuckle. "An American expression," he explained. "Maybe it's Norwegian also—to get attention.

"Anyway," he continued, "when I heard you mention printing, Kari, I wondered if your company might have been in business long enough to have printed Greta's poster that was made into a will."

"Oh yes," said Kari, "our company goes back to about the start of the century, and it was taken over by the Nazis during the war. Among other projects, we publish law books, and I work in that department."

"That makes me curious about whether the Greta's poster might have been produced at your printing plant," Harold added, with a hopeful expression on his face. "Maybe you would be willing

to look for some records of the wartime printing activity. We all know that the Germans kept precise records of everything they did, so there's a chance of some reference to that poster."

"I'll start searching right away," volunteered Kari. "Maybe I can find some information that would be helpful to your family, Karl."

After that comment from Kari, Karl reached around her to give her a firm hug, to go with his firm "thank you!"

"That reminds me, Harold," said Karl. "You said that you haven't seen one of the German gun emplacements on our island of Karmøy."

"Not close-up, anyway," said Karl, "just from the ship or overhead from the helicopter. Even then, the one I saw looked huge and impressive. And I experienced a strange reaction about that gun emplacement—that the design with its arched dome and curved wing looked rather distinctive and appealing."

"What a coincidence!" exclaimed Karl. "That's much like what Greta said.

"You may not know it, but her Col. Schorn designed that structure and supervised the building of it," continued Karl.

"Well, all this will give us plenty to think about," concluded Sofie.

"It certainly will," commented Karl. "And with that, I think we should say 'good night,' till we meet again."

He and Kari both hugged Sofie, and Kari hugged Harold after Harold shook hands with Karl.

"Well, I still want to see your photo gallery, if you're willing, Sofie," said Harold.

"Naturally, I'm curious about the one that's face-down," he smiled.

"Just an on-again, off-again boyfriend from Trondheim," Sofie answered, with slight embarrassment.

Then, as she began to identify and explain other photos, she began to cry softly as she pointed out an older couple.

"My Stensgard grandparents," she explained. "Grandfather was executed by the Gestapo near the end of the war for his resistance activity."

Harold put his arms around her as she leaned against him, still crying.

"Sorry," she apologized, "I get all weepy when I look at that photo. Should turn it around so I don't see it so often."

Harold pulled her close, kissed her forehead, then the tip of her nose…then, after a brief hesitation, kissed her on her lips.

She responded by wrapping her arms around him, and returning his kiss.

Then they walked together to the door. "Hate to say good night…but just till another time soon, I hope," he said softly.

"Yes, let's do so again, soon," she smiled. "We definitely need to meet for more planning."

As he started out the door, she asked: "Would you like to go on a picnic? We could take the ferry to Karmøy. I could show you the gun emplacement, where I've visited several times."

"Perfect way…place…and person for conducting research," he affirmed, with a happy smile.

"Maybe we could make a foursome again," she said.

"Good idea," he nodded. "Fun…and sharing of reactions and discoveries could be valuable."

"I'll start the picnic rolling with Kari," she added, as she hugged and kissed him again.

But before that excursion, they got together again to focus on the presentation to the community, aimed at increasing the mutual understanding of the petroleum industry personnel and the Norwegian residents of Stavanger.

Harold shared his news of a skilled presentation helper at his office, "and he's on standby, waiting to put our ideas into an interesting form for our slide presentation.

"So," he urged, "before the night is over, I hope we can identify the key points we want to make, so we can prepare titles and graphics to add emphasis and information."

With that request, the four buckled down and outlined the basics of the presentation.

"I can take it from here," said Harold, "with some help at the office. Then we can divide the responsibility for delivering the message during the meeting.

"We can prepare slides with data from the city and from the petroleum companies. Charts can help show trends, which would also add impact."

"It's our idea," said Karl, "so I guess it's up to the four of us to take the lead. Now that the city's population is up to about 50,000, we will no doubt fill the school auditorium."

"We'll soon have a flyer ready to distribute," said Kari. "And we can distribute through churches, businesses and local government offices."

"And schools, of course, where it started," added Sofie, with a smile.

"I think we can share some important and interesting information," commented Harold, "but just the idea of the gathering might be as important as the actual information."

"We should have no trouble lining up appropriate community and industry representatives to be available to answer specific questions," Karl explained.

"A social time, with refreshments, after the formal presentation and Q and A session, should encourage further interaction and information exchange," added Kari.

"I think local stores and bakeries will be willing to donate, and high school students probably would be glad to volunteer as servers. And help clean up afterward," she smiled.

Action proved a bit more challenging than talk, but with delegation and good follow-through, the presentation came off nearly as they hoped.

Immediate positive response from many in the audience proved rewarding…and satisfying. And encouraging about a repeat performance.

So, after the dust settled, all involved—or almost all anyway—considered the gathering an important and enjoyable time of learning and sharing.

And the quartet of planners, Karl, Kari, Harold and Sofie, celebrated later at Sofie's apartment with a toast of wine to the success of the initial event.

And they pondered the question of whether or not this kind of gathering should become a periodic event in the future.

Chapter 7: Picnic with Purpose

On a sunny Saturday mid-morning, a Volkswagen two-door sedan pulled up at Sofie's apartment building. Harold hopped out of the passenger's side and bounded up the stairs to the second floor.

In the "beetle," Karl in the driver's seat looked back at Kari with a grin: "I'd say he's eager to see Sofie!"

"Yes," Kari chuckled, "I think he's not just hungry for what he expects in the picnic basket. Sofie confessed to me while we were planning this outing that she and Harold seem to be 'hitting it off,' as the kids say in their international slang."

"That's great!" Karl reacted. "I like them both, and look forward to their company again. And follow their lead with international smooching," he added, with a slight smile.

"Yes, I bet you will…as if you need any encouragement," said Kari, with her own hint of a smile.

When Harold and Sofie came toward the car, holding hands while he carried her picnic basket, he grinned like a Cheshire cat. And she smiled demurely as she climbed in the back, while Karl and Harold placed her picnic basket in the trunk at the front.

"Sorry to put you in the back, Sofie," apologized Karl, "but I thought Harold's long legs might fit better in the front seat. Tight fit at that."

"The students should see their 'John Wayne' now," laughed Sofie, "scrunched up in like a bug in a bug. Sorry…no horse for you to ride into the sunset today."

"I guess I should be back in the saddle again with the wide open spaces for my long legs," laughed Harold. "I'd decided that if

John Wayne could switch from football to horseback, I could too."

Later, he thanked Karl for heading West toward the Atlantic.

"Weren't the Norwegian Vikings always heading West!" declared Harold. "Even in Seattle, our statue of Leif Erikson looks West, eager to go across the Pacific. After all, the Vikings had already crossed the Atlantic."

"Soon we'll be on the ferry for the ride to Skudesneshavn, sort of going west," laughed Karl. "Then driving west through the island for a dramatic view of the North Sea. And an even more dramatic view of the massive German bunker!. We can park in a pasture and walk a short distance to have our picnic on a flat concrete area near the entrance. I have to admit that it will seem strange, even though I've been here several times…but not for a picnic."

"The grassy area tends to be damp from the usual moisture off the ocean," explained Kari. "Probably not like your West Coast."

"Well, the southwest coast in California compares to the Riviera, but where I'm from, in Washington State, we get lots of rain and low clouds. So I feel at home here. But when the sky clears at home, we do get dramatic views of snow-capped mountains.

"Sounds mighty appealing to a Norwegian," said a wistful Sofie. "Hope to see it someday."

Mount Rainier in Washington State

"It would be a great trip for our

foursome," said Harold. "Let's plan for a picnic at Mount Rainier!"

"Is it big?" asked Kari.

"Second highest in the main part of the United States," answered Harold. "About 14,000 feet."

"That's more than 4,000 meters," Karl figured.

"Wow!" exclaimed Kari.

"Have you climbed to the top?" asked Sofie.

"Once, when our group of climbers could hike around glaciers and over the snow in the summer. Breathtaking…because you might say that climb definitely left me out of breath," Harold laughed at his own play on words. "And once was enough. Just to prove I could do it."

"I think the students will be proud to hear that…as I am," said Sofie.

After spreading a tarp to protect them from the moisture on the concrete slab, Kari and Sofie opened their picnic baskets to set out food.

"We don't need fancy dinnerware," declared Kari, because this is picnic style, with just some paper plates and plastic spoons, forks and knives. "Besides, we've learned that you Americans just say that 'fingers were made before forks'."

"Ta-da! Here's our special of the day—hot dogs—a combination of American and Norwegian, with pork sausages and whole-wheat buns, relish, catsup…but no onions, because their smell might overcome us inside the Volkswagen," laughed Sofie.

"And we have potato salad, pickles and even some potato chips," Kari announced as she laid out the array.

"Just like home," said Harold, "though I'll bet your 'Viking' sausages will beat the quality of our hot dogs."

"And lemonade to drink," said Karl. "Another American favorite, and growing in popularity in Norway," said Sofie.

"And some home-made chocolate cake, by Sofie, to top it off," Kari smiled in appreciation. "Appropriately, German chocolate, I presume."

Then their talk drifted to the scope, shape and purpose of the bunker.

"I thought of an interesting aspect of the design," Harold said thoughtfully. "The shape of the bunker—with the arc of the smaller dome flowing into the arc of the upper and larger dome—reminds me, Karl, of the shape of your Volkswagen. It too has the low arc of the trunk in front merging into the arc of the rest of the body."

"What an interesting comparison!" Sofie stated. "And I do see that pattern of arcs."

"Well, I have wondered about the thinking of that noted designer of the Volkswagen, the car of the people," pondered Karl. "Dr. Ferdinand Porsche designed the 'beetle' in response to Hitler. And I wonder why. Just as I wonder about the commitment to this kind of bunker design by Col. Schorn."

"Now that you mention such apparent contradictions," Harold interjected, "I do wonder why you as a Norwegian would buy a German car, after how your people suffered from the Nazis?"

"Much of that feeling has faded after a quarter-century," Karl replied, hesitantly and thoughtfully. "I suppose now it comes down to high quality and low price."

"Maybe your aunt Greta will help us understand," responded Harold.

"Yes, I think an opinion from Greta could be revealing," said a solemn Karl. "Maybe I will gain more understanding of her as well."

Later, before leaving the location, they looked inside the bunker. Even in the semi-darkness, they could look out the wide "windshield" to take in a panoramic view of the North Sea.

Karl's flashlight revealed the basic space inside and the compartments for special purposes. Wiring led to metal boxes connecting electrical circuits.

"I do have to admit," mumbled Karl, "that the Germans planned well and built well too."

"Before we leave this island," suggested Kari as they packed to leave, "I think we should take this American tourist on a tour."

"Good idea!" said Karl, so they squeezed back into the 'bug.'

As Karl headed north, Harold admired the pastoral scenery.

"Here's Hillesland Lake," announced Karl. "It stretches out quite a distance…shallow but deep enough for our waterskiing."

"Sounds like fun," declared Harold.

And the other three voted a vigorous "yes!"

Farther on, Kari pointed out the Avaldsnes church. "During the war, it made an appealing shooting target for planes of both sides. So our Norwegian patriots decided to protect it with camouflage of foliage and netting on top of a wooden framework so it looked just like a small hill.

"Now," she continued, "the history organization has made this into a significant site. And that same group is creating a Viking village nearby, with plans for a Viking museum as well."

"Now, Harold, we've saved the best for last," said Karl, as he drove farther on the island. Then he pointed a large brick structure. "This," he announced, "is the smelter that produced the copper that became the skin of your Statue of Liberty."

Statue of Liberty

"Wow! What an amazing surprise!" exclaimed Harold. "I had no idea where the Statue's copper skin came from. I thought France…but this makes her all the more appealing. Wonderful!"

"That makes a great finale for today," concluded Karl. "Another time, we might take in a museum in Haugesund, which offers sad memories of the stressful life here during the war. It shows how desperate the people were for food during the war. Fortunately, the sea helped their survival, because fishing continued, though the Germans took the bulk of the daily catch. One museum even shows shoes made from fish skins. That's how desperate they were."

A somber tone prevailed as Karl drove back to the ferry and then on into Stavanger.

At Sofie's place, Karl offered to drive Harold to his apartment.

"No," he declined, "I can walk from here, after I help Sofie with her basket."

Kari smiled as the Volkswagen "musical chairs" occurred again. First Harold got out, then Sofie exited, next Kari got out, hugged Sofie and Harold, and then got in the passenger's seat. Karl came around the front of the Volkswagen for a hug and handshake, and they all exchanged pleasantries about the picnic.

And Harold added his thanks for a revealing and enjoyable experience.

At Sofie's door, Harold set the picnic basket down and put his arms around Sofie.

"Thanks for helping make this possible," he said, before kissing her.

"I certainly enjoy having a lovely teacher helping me learn," he said.

"Would you like to come in? I can make coffee and "rustle up" some supper," invited Sofie.

"The simple answer is 'yes, I'd like to come in.' But the safer answer is 'not tonight.' Too tempting," he said.

"I know," she said. "So thanks for a wonderful day, even if a bit somber at times."

"Learning more about a source of our Statue of Liberty, right here in Norway, sure topped off the tour in grand style!" Harold declared.

"Reminds me of the excitement of seeing it in the New York harbor."

"Maybe someday for me too," said Sofie.

"Well, for now, let me see what celebration I can cook up to honor the Statue of Liberty," suggested Harold.

Chapter 8: 4th of July

"I am sorry to say," announced Harold at a gathering of family and friends at the Onstads', "that Norway doesn't honor the American 4th of July by giving the day off. But I think we can celebrate our Independence Day at night, when all we invite might have a chance to participate.

"So, I, with the help of our company, have booked a harbor cruise boat to handle about 50 guests. You folks, plus colleagues, member of the International Church, and maybe some Stavanger dignitaries.

"We'll have food and drinks and music," he added, "but probably not enough room for dancing, sorry to say."

"Wow! That sounds great," shouted Jakob. "At least, I hope I can come too!"

"Sure can, and it should be great!" echoed Harold.

Then "Great!" rang out in various tones from others.

"At least the cruise will float by for some precision-timed fireworks from shore," added Harold, "and the combo will play the *Stars & Stripes Forever.*

"But, now for our trivia test," he laughed. "Who wrote that song?"

"John Philip Sousa!" shouted Sofie. "Our students love it!"

"Next question," announced Harold, standing tall and looking presidential. "What American president was of Norwegian descent?"

When no one ventured a guess, Harold hinted: "The Father of Our Country."

"George Washington," shouted Kari.

"I didn't know that," muttered Karl. "But it's good to know!"

"Well, I know about Marilyn Monroe," shouted Jakob. "Her father came from Norway."

"Definitely good to know," laughed Haaken. "Looks just like Olga…Marilyn, that is, not her father."

"Good thing you made that clear," chuckled Olga.

"Okay," announced Harold, "one final…and easy… question. What current American leader has roots in Norway…even a town in Norway with his name?"

"Vice President Mondale!" exclaimed Kari.

"Good for you, Kari!" commended Harold. "You've named a president and a vice president connected with Norway. So you deserve this silver dollar!" as he flipped it up for her to catch.

"That eagle reminds me of the masthead of a Viking ship. Maybe that's what we should have for our cruise. Or were they all left by Leif in America!"

"Well, in America, how do you celebrate your Independence Day?" asked Karl.

"A bit more than what we'll do," explained Harold, "with fireworks, boats cruising in the Puget Sound, planes flying overhead, baseball games, political speeches, even acknowledgement in churches, and parades with marching bands…playing the music of John Philip Sousa, of course.

"And with American flags flying everywhere.

"Great fun…except for burned fingers like I got one year from firecrackers…and an upset stomach from too many hot dogs and too much potato salad.

"Well, in America's Last Frontier in the West, some do get a bit out of hand and end up in the *calaboose* from drinking and fighting."

"What's a calaboose? Another term from the West?" asked Jakob.

"Just a fancy French term meaning 'jail,'" chuckled Harold. "So watch your drinking and fighting!"

Then, like the announcement about the recent community meeting, news of the 4th of July cruise spread in Stavanger. Not that residents expected to get on board, but many wanted to share in the ceremony by at least waving flags and shouting greetings as the boat cruised by.

After the 4th came and went, Harold announced simply to all the passengers: "That was the best Independence Day I ever had. Thanks…and I salute you."

Chapter 9: Shift to Sweden

Because of the petroleum industry, small commercial planes flew in and out of Stavanger at a steady and frequent pace.

So Karl picked up Harold and they headed for the airport in his "bug."

"I'll just leave the car at the airport," explained Karl, "because I imagine we won't be in Sweden for long. Glad I learned that a Statoil plane was scheduled to take an executive to Kristianstad for a meeting at the university over this week.

"After going by helicopter many times to the oil platform, I got to know some of the flight staff with Statoil, so I got wind of this flight. Space to spare, so it was easy to arrange for a hop."

"Smart thinking," said Harold. "Faster. Easy on the pocketbook too," he laughed.

After they boarded the Learjet and it took off a short time later, Karl summarized the schedule: "We should arrive late in the afternoon at Kristianstad, and Greta and Hanna will pick us up at the airport there. So I count on an efficient, enjoyable and productive visit with them.

"They're favorites of mine," Karl continued. "With you helping with basic questions, I hope I can gain a better understanding of their past and present.

"Then if we have a chance to see that special poster given to Greta, we might be able to link that to the records at the printing plant where Kari works."

"What can you tell me about them in advance?" asked Harold.

"Both very attractive and well educated," said Karl as he started describing them. "Greta must be about 60 and retired as a teacher. Never married, except, you might say, to Col. Schorn.

"I assume she will know a lot more about him, and, depending on what she wants to tell, so we could get some leads for research about him in his area in Germany.

"I guess Hanna is a bit older than I. About 32. Has a doctor's degree in economics and teaches at Kristianstad College. Not married, as far as I know. Guess we can find out for sure."

Kristianstad College in southern Sweden, where Dr. Hanna Onstad teaches economics.

"Now we've got a quite a way to go, so I think I'll try to snooze a while," concluded Karl.

As Harold pondered this current situation, he thought about his interest in Sofie, about his past and future, and he didn't feel at all sleepy. But he felt eager to learn more about the mysterious poster—also being considered as a "will." But for what?

After a soft landing at the small airport in Kristianstad, Karl spotted his relatives at the gate.

He waved at them and waved for Harold to follow him in the direction he pointed.

Obviously mother and daughter, and both very attractive, thought Harold. The tall, slim younger woman looked handsome in her charcoal business suit, and her mother also looked lovely in her colorful casual attire. Then he reminded himself of their names, Greta, the mother, and Hanna, her daughter, both by the name of Onstad.

Both hugged Karl, and then he introduced them to Harold. After they shook hands, Greta grabbed Karl by his arm and guided him toward a waiting car in the parking area.

So Harold politely held Hanna's arm and they followed the lead of Karl and Greta.

"Glad to have an American in our clan," announced Hanna. "Or at least as a good friend of Karl. He said that you have Swedish roots too. So much the better," she smiled as she looked up at Harold.

"You sound American too," said Harold, with an inquisitive smile. "Or British."

"Just a Swedish lilt in my English," she said. "But not in yours."

"No, not even from our Scandinavian area of Seattle," answered Harold. "Guess my great-grandparents and everyone else in the family were glad to leave the tough times of Sweden behind."

"I can understand that," said Hanna. "Karl may have told you that my mother was glad to leave Norway behind too."

"So I heard, and that's one reason we're here. Karl wants to clear the air for his family, as much as possible. And I'm curious too, though it's really none of my business. So if I sound too inquisitive, just let me know."

"I will. I do that sometimes in my college classes, though Swedish students tend to be diligent and courteous," she

explained. "Can't say that for a lot of the new immigrants, though.

"Well, here we are," she continued. "Our limousine awaits us. My SAAB has an American link also, with your General Motors."

"You'd better sit in front with Hanna," said Greta, with a restrained smile. "With your longs legs, you must be a basketball player."

"I'd better answer that," said Karl. "Harold would play down his starring role with the University of Washington football team."

"Sounds like Swedish restraint," laughed Hanna.

"I saw you in a video clip when you were honored after a Rose Bowl game, Harold, so I thought I'd better brag a bit for you," chuckled Karl.

"Naturally," he said to Greta and Hanna, "he's not the bragging type."

"I think I might have seen that film clip too," laughed Hanna. "Just tuned in to see some of those football 'hunks'. Football is not big on our campus. I have noticed that all you American football players don't look alike, at least when you take your helmets off."

"Part of the melting pot we talk about," explained Harold. "Though I think of our mix more like a salad, where the different ingredients can still be identified."

"Does that suit you?" asked Greta, from the back seat.

"Sure does!" answered Harold. "That's our promise and our hope, though the pace of change seems to drag on more slowly than expected.

"You probably know that from the research of our Negro population by your Gunnar Myrdal," Harold added.

"Famous in Sweden," said Hanna, "but the town by that name is in Norway. So you see, we Scandinavians do share."

"We've been talking about that challenge," said Karl. "Harold seems to know a lot about our divisive history, including the efforts by his royal namesake to unify a splintered Norway."

"That brings up a thought," said Hanna. "When I heard about your visit, I thought a presentation by the two of you to my economics class would be desirable. Offer international insight about petroleum economics—and respond to whatever my class might be curious about.

"So, if you're willing, I could schedule you to meet with my class on Wednesday. That would give me time to prime my students to be ready with some questions."

"And for us to get our act together," said Karl.

"We're not due to return to Norway till Friday," added Karl. "So Wednesday should fit. And I'd sure like to be part of that kind of discussion. The student reactions might even add another angle to a community presentation in Stavanger that four of us are planning—as it relates to acceptance of change.

"And we should still have time for Harold to connect with some of his Swedish relatives he got to know when they visited America," added Karl.

"They live in Broby," said Harold, "and I met some of them when they visited Seattle."

"The town's not very big, and not far away at all," said Greta, "and I can take you there. It would be interesting to meet your Swedish relatives."

"Here we are at my place," said Hanna. "Just put your thoughts on 'pause,' so we can pick up on this conversation during dinner.

"At least I assume you might be hungry by now," she added.

"And after dinner, we'll drive out to Mother's farm," Hanna explained. "She has a lot more room than I have in my apartment, so you will be able to kick back and sleep in comfort."

"Mother and I prepared goose for dinner," said Hanna. "Meant for special occasions, like this," she smiled.

"Well, this gives me a new acquaintance with a goose," joked Harold.

"I only know about the goose who flew Nils Holgersson all over Sweden," he laughed. "That story gave me a good picture of Sweden when I was a kid.

"And now I look forward to this delicious Swedish food favorite too. So thanks in advance for your gracious hospitality."

As their lively discussion proceeded, Greta noted that this didn't seem like a typical serene Swedish 'supper'.

"More like a meal with Texas oilmen," she laughed.

During a lapse in the conversation, Karl introduced a different topic.

"Harold has worked in the printing business in Seattle," Karl commented cautiously, "and he's curious about your marked-up wartime German poster that also seems to have become a legal document."

Greta began explaining: "I imagine, Harold, that you know about my 'de facto' husband, Col. Ludwig Schorn."

"Yes," said Harold, "I learned that he was an architect, engineer and printer, and that he distributed German propaganda posters in Norway as part of his responsibility. I understand that the poster he gave you was written on to become a 'will'. Is that correct?"

Greta declared: "It definitely is!"

Harold continued, "Karl's girlfriend Kari works in the oldest printing company in Stavanger, so we asked her to search for any link in the company to those posters. If we find anything, we'll let you know, though I don't know what bearing it might have on your particular poster," said Harold.

"At the farm, I can show you my cherished poster so you and Karl can see its significance," said Greta. "And I will wait eagerly for any history you can find about the printing of the poster.

"Of course, the real puzzle relates to the implications of the wording Ludwig added by pen. Or someone did. And what kind of 'valuable' did he allude to? And if it exists, where might it be concealed?"

Then Hanna offered "her take" on the "treasure": "Maybe whatever he referred to was found and taken long ago. But we still think of this document as an act of love from my father. That in itself is a treasure," she added, as both women teared up.

"You'll soon see that cherished and protected poster," said Hanna. "Of course, we have copied it for reference, so we could provide you a copy for your research with the printing company in Stavanger."

Then she smiled: "Mother has promised her renowned coffee and dessert to finish our dinner. So climb aboard again in the SAAB for a drive in the country."

In a short time, Hanna turned down a lane that went to a rustic and rambling farm house, with assorted outbuildings.

After Greta climbed out of her back seat, she announced: "Haul your luggage in, and Hanna can show you your rooms while I put the coffee on and warm up the crepes."

"Actually, that's just her fancy name for lefse," explained Hanna, "but the taste is great. After all, what's in a name!"

As they sipped coffee and savored the lefse, Hanna showed a Xerox copy of the inscribed poster. "We locked the real thing in our safe—so it's protected from fire as well as theft.

"Anyway, here it is for your perusal," said Hanna. "We don't know if it was produced in Germany and brought to Norway, either already printed or already set in type to be printed in Stavanger."

"As I recall from the printing company I worked for in Seattle," said Harold, "an advertiser would have a papier mache mat made of a printing form to send to us for casting a plate for printing. A lot lighter and easier to send than a heavy printing plate.

"Then we would suspend that mat in a casting case and pour in molten lead to make a printing plate. Your husband no doubt knew of that process too, so the printer in Germany could have made lead printing plates for him to bring along...or did he just bring a papier mache impression of the poster's type arrangement to be cast as a printing plate in Stavanger?"

"I do vaguely recall one time when he mentioned how heavy the printing plates were when he brought them along from

Germany," explained Greta. "He said he had explained that printing plates were needed for possible use by several printers

Casting molten metal in a form to make printing plates.

at various other locations. Perhaps the plates are gathering dust at the Stavanger printing plant."

"With luck, maybe Kari will track down that plate or plates," suggested Karl.

"Ordinarily, printers would melt down the lead plates after they had served their purpose," Harold mentioned. "But the thorough Germans may have preferred to store them. We'll see—maybe.

"Anyway, this document certainly rivets a person's attention, with all those signatures, including your husband's."

Then Harold posed a question: "Are you familiar with a mystery written by Edgar Allan Poe, called *The Gold Bug*?"

"I'm acquainted with Poe and the problem-solving in his stories, but not with that particular one," said Hanna.

"*The Gold Bug* includes complicated directions toward a treasure, so I wonder if the wording in your poster might have some allusions to hiding something…what and where," pondered Harold.

"We've read the wording on the poster backward and forward," said Greta, "but the words don't reveal anything, as far as we can determine.

"But your reference to *The Gold Bug* story reminds me that Ludwig said he had been worried about protecting his family's investment in gold from the Nazis. So maybe we should be looking for a clue about that in the wording."

"We'll be looking too," said Karl, "if you can provide us a copy of the annotated poster."

"Sure can," said Greta.

"Incidentally," continued Karl, "Harold has wondered, Greta, if it doesn't embarrass or anger you or Hanna, how you became involved with a German officer—when that was so offensive to most Norwegians. Maybe some more information from you would help with the mystery."

"Okay," Greta responded, with hesitation, "I got accustomed to seeing Ludwig in our community, where he had an office next to where I taught. First, I'll admit he was handsome. And friendly. At school, he would glance through the glass in the door to see the students practicing art, with me as their mentor.

"One day I guess he noticed I had messed up in helping an older student draw a building. We were laughing about how we had distorted it.

"So he dropped in to literally lend a hand. He grinned when he saw our misshapen structure, so that began his instruction about dealing with perspective of buildings…and of various objects too.

"I felt his companionship to be enjoyable…and he didn't seem like a typical German soldier. Actually, he wasn't typical. He was a skilled architect, reluctantly doing his duty, which included planning and supervising the installation of gun emplacements on the coast of Europe.

"Well, a romance developed between us…and eventually led to the birth of Hanna. And that caused attacks and ostracism by people in our community.

"Then the German high command sent Ludwig south to help fortify the coastal defenses along the English Channel. We learned that he died there during the Allied invasion. And we heard that the Gestapo had already eliminated his family because of their resistance to Hitler.

"Finally, in desperation, I took my baby and fled to Sweden, as Ludwig had arranged earlier."

"Thanks for the explanation," said Karl. "And I'm so relieved to see how happy and settled you are in Sweden.

"But is there any romance in your life now, Greta?" he asked.

"Both of us seem to fit the pattern of Sweden," Hanna responded. "Stay private and wait for relationships that might develop. Meantime, Mother enjoys retirement from her career in Sweden and I enjoy the challenge of learning and teaching.

"Whatever will be, will be…" she added, without a sign of bitterness.

"Her favorite song when she was growing up," said Greta.

"… the future is ours to see!" added Harold.

"Okay," continued Hanna, "now it's time for me to ask a personal question. What is your situation, Harold?"

"Norway interfered in my personal life," said Harold, with a restrained laugh. "I was engaged to a wonderful woman in Seattle, but then she decided she could not leave her career as a lawyer and move to Norway. So, she gave back the engagement ring.

"After I boarded the SAS plane for Copenhagen, the challenges in my new assignment engulfed me—figuratively and literally.

"Now I am discovering charming female companionship in Scandinavia, as well as friendship with the others in my new community—at work, in the church and at the school where Sofie teaches.

"So patience prevails," smiled Harold.

Then Harold shifted focus to his copy of the poster given to him by Greta.

> **NORSE**
> **join our**
> **German**
> **battle**
> **against**
> **Communism**
> **!**

"Karl, would you read the poster aloud?" asked Harold. "You can deal with the Norwegian better than I can, and maybe we can hear a helpful nuance in the wording."

Slowly, with emphasis on key words, Karl read the large letters of the vertical 12" x 18" poster.

One by one, they all admitted that they saw no clues in that simple message.

But as Harold studied his copy again, he noticed some curved lines where Col. Schorn had placed his signature to attest to the assignment of his assets to Greta Onstad.

He pointed and asked about the two delicate arcs marked there: "Are they on the original poster, or are they just marks that are the result of copying?"

Greta checked and pointed at the marks. "Don't know why I didn't notice those before," she said solemnly. "I'm sure they were not on the printed original. Guess my emotions blurred my sight."

"Then, do you know any reason he might have added those subtle arcs below his name?" asked Harold.

"Probably just a flourish to emphasize the importance of his signature," guessed Greta. "As I said, I hadn't even noticed before."

"Sharp eyes for noticing that, Harold," complimented Hanna.

"Comes from checking and proofreading lots of printed pieces in my previous life," said Harold. "Caution motivated by fear—that I'd miss a mistake. Enough to drive a person crazy sometimes."

Then Harold blurted suddenly: "Remember, Karl, when I said the lines of your Volkswagen were similar to the gun emplacement on Karmøy. Maybe those curves added by Col. Schorn refer to the shape of the bunker. And to a location of Greta's hidden treasure!"

"Good thinking, Mr. Poe!" said Hanna.

Norwegian Coast along North Sea

Concrete base for artillery turret

"That opens up a challenge," said Karl. "It could start us on an exhaustive search, like finding a needle in a haystack, as you Americans say. But I think we should get on it right away when we return."

"We can start by analyzing the structure carefully, to see if there are some obvious places for hiding something—of a size we don't even know," grimaced Harold, as he thought of the scope of the search.

"I do recall metal junction boxes installed in the concrete for the electrical power lines and fixtures," said Karl. "Could be too obvious, but a relatively easy start of a search.

"The bunker was the control point for the artillery turret in that massive ring below the dome. I hope the artillery wasn't a hiding place, because it was hauled away. Not likely," Karl assured himself. "But that concrete ring for the turret might have been accessible for burying an item at the time of construction."

"Guess we could start by checking the walls and floor of the dome, hammering steadily in order to listen for a hollow sound," said Harold. "And if we do hear that, we'd have to break through the concrete for access.

"Any problem from the Norwegian officials in doing that?"

"I'll ask Kari if she will check that out," said Karl. "Red tape, as you Americans refer to government regulations, could cause a delay at the least and disapproval at the worst."

"Nothing like throwing cold water on a hope," lamented Harold. "But now I suppose we should try for some sleep. We have to rest up to look alert and appear to be knowledgeable as Hanna's guest speakers."

"Let's think of this as another rehearsal for our Stavanger community presentation—and as practice for a possible repeat performance again later in Stavanger," Karl commented.

"I do admit the excitement of our Poe deductions has left me a bit tense," said Harold, "and a good night's sleep is overdue."

"Oh, I think our tough Navy Seal needs a Swedish massage," laughed Hanna, as she squeezed Harold's shoulders from behind.

"Getting sleepy now?" she asked.

"Oh yes, this is wonderfully relaxing," replied Harold.

"Don't relax too much," cautioned Hanna. "You have to brace yourself for the questions I have in mind for the class. I do tend to be a bit critical of Americans' use of resources."

"Beware, Harold," chuckled Greta. "She'll egg the students on as well."

"With that warning, I think I will say 'good night' and head for my home in the city," grinned Hanna. "See you Wednesday, if not sooner."

Chapter 10: Swedish Clan

With time to spare before their appearance in Hanna's University class, Harold and Karl rode along with Greta to Broby to visit Harold's Swedish relatives.

During the short drive through farmland, Harold explained that he had met the Olssons on his mother's side of the family when two of the college students, Tina and Anja, visited Seattle several years ago.

"They're now living in other parts of Sweden," Harold explained, "but in my phone conversation with their parents, Emil and Edith, they said their daughters and families are hoping to join us at the family farm."

At the Broby farm, bold blocks of colors startled them as they

approached the entrance to a tree-lined *alley*.

"That sign makes me feel welcome. All it needs is a greeting in Swedish," said Harold, "and…I guess Finnish as well."

"Sure looks like the influence of Finland," said Greta, "not the usual Swedish somberness. Well, considering what Finland has endured—battered endlessly by Germany and Russia during the war—they have good reason to burst out with strong shapes and colors. That's reflected in some of their music too."

"I admire Finland for another reason," reflected Karl. "For their *sisu,* to never give up! Maybe their actions in fending off the Soviet Union inspired the development of NATO…the North Atlantic Treaty Organization…to contain the Russians. At that time, 20 years ago and well before our oil, the Soviets even had their eye on Norway. Perhaps for the link to the Atlantic."

"All the more reason to declare that the banner looks great!" commented Harold. "In a positive way, it even reminds me of the contemporary IKEA type of design."

"Sure does," echoed Karl.

"Well, IKEA started not far from here in Smaland," explained Greta, "where the poor soil pushed residents past farming and into other ventures. Probably the same for Finland."

When Greta parked in the gravel courtyard, a large group, young and old, came out of the house to meet them.

"Great to see the two I recognize," said Harold, "and they still look as lovely as I remember."

"With their parents and their families, I presume."

As they got out of the car, Harold got hugs from Tina and Anja.

"Our movie star look-alike!" Tina exclaimed. "You look even better now!"

"And so do you…Tina," replied Harold.

"Right, and I'm Anja," said the other.

"Both still lovely to look at," said Harold, "and as friendly as when I met you in Seattle."

Then Harold introduced Greta and Karl, as Tina and Anja introduced their parents, husbands and the four young children.

"Time for *fika*," invited Edith. "I believe you Americans call it a coffee break. Anyway, a good time to get acquainted."

"And good sweet rolls," said little Anders.

"We admire your farm…the alley in the rows of trees…and the dramatic blocks of color," Harold said, to start the conversation.

"And we particularly enjoy the graphic design and glass created in Finland, now that we're retired with enough leisure time to appreciate design," said Emil.

"Look at that cabinet, full of glass," said Claus, husband of Tina. "Quite a collection, including some of their own creations."

"Do you have your own glass shop?" asked Greta.

"Too complicated and expensive for us," replied Emil, "so we go to a nearby shop to create our glass. And we welcome the expert assistance there, too. Great ongoing education!"

"We've learned that your community in Washington produces some outstanding glass," said Anja. "Our folks frequently mention that as inspiration."

"Dale Chihuly leads the way in our Puget Sound area as he experiments with a free-form approach with glass," Harold explained. "And, by the way, his mother is Swedish-American. So I think we can take special pride in his creativity."

"We enjoy screen-printing too," explained Edith. "We have space for that, and it doesn't require quite the amount of equipment and materials as glass-blowing. Good on fabrics, as well as for items such as our sign by the entrance to our alley."

"Looks like you use wood effectively, too," Karl admired.

"Also part of our Scandinavian craft tradition," said Emil. "And that reflects the achievements in Finnish architecture as well."

"Not tools of the past, though," smiled Arne, husband of Anja. "Maybe they'll take you on a tour of the wood shop, with the latest of power equipment. Even I get to use it once in a while."

Dale Chihuly glass

"From your infrequent letters, Harold, I gather that you've taken on a lot of challenges since we met you," said Tina, "including serving with the Navy Seals. We've heard about their exploits."

"Grit and guts, we sometimes called it," answered Harold. "Then I had to go underwater for a different purpose in the North Sea, checking on the structure of the oil platforms. So I can happily do without both of those assignments now and do my oil-marketing analyses from the comfort of my desk in Stavanger."

"How about you, Karl?" asked Emil. "You in the oil business too?"

"On the production side," said Karl, "trying to make sure the oil keeps flowing to the markets Harold connects with."

"And, I understand," interjected Edith, "that you're a farmer too, Greta."

"Maybe like your farming," said Greta, "if you have a small plot of land like I do. I experiment with plants, mostly flowers, but some vegetables. But I sure wish our growing season were longer to give my experiments a better chance."

"Now that you're living here, Harold, what impresses you about Scandinavia?" asked Tina.

"Mostly, I've learned that Norway reveals evolving contradictions," Harold mentioned. "Norwegians seem to resist change by clinging to territorial traditions in clothing and décor. But war forced change and set the stage for the leadership of Trygve Lie in the new United Nations. Now their oil boom has forced major changes as they adapt to us foreigners. They criticize American materialism, but they seem to be enjoying living high on the hog themselves now."

"I'm afraid he hit the nail on the head," Karl admitted.

"Still, once you get beyond their Xenophobia," added Harold, "they're friendly and trustworthy, honest and generous. Much like my perception of the rest of Scandinavia.

"And the women are beautiful and beguiling," Harold added, with a warm smile for Greta and the Olsson women.

"Don't be beguiled tomorrow," warned Greta, "when you and Karl address Hanna's economics class. I think she might serve up some challenging

Trygve Lie

81

questions for you.

"For your information," said Karl, "my cousin Hanna teaches economics at Kristianstad College, and she has invited the two of us to talk about the oil business and its impact internationally. We don't know what to expect…but to prepare for the worst from the students…and Hanna," he added with a smile.

"From my experience," said Tina, "I found Americans to be friendly, eager to learn, respectful of others…actually, not unlike Sweden. Well, discounting the phoniness of Hollywood. And the ethnic clashes…which we are now getting a taste of here."

"Ditto," said Anja. "We were impressed…amazed…by what Americans have accomplished, knowing that many of them left here virtually destitute a hundred years ago."

"Thanks to both of you for a good pep talk," smiled Harold. "And your welcome insights to share about America."

Later, on the way back to Greta's farm after a tour and further food from the Olssons, the trio concurred that Emil and Edith had developed a good life and benefited from their loving family.

"In looking back, I envy them," said Greta. "But life in Sweden has been different from my life in Norway. Still, I am thankful for what I've salvaged. For Hanna. And for relatives such as you, Karl. And friends such as you, Harold."

When they arrived at Greta's farm, Hanna greeted them…with coffee and lefse…"just the way Mother taught me," she smiled.

After the trio shared farm impressions, Hanna warned Karl and Harold that her students had identified issues to mention tomorrow.

"We called our issues 'Jeopardy!' just to keep you on your toes," reported Hanna, with a wry grin.

"Tell me about Jeopardy," requested Greta.

"That one's for you, Harold," teased Karl.

Harold did his best to explain: "It's a TV quiz show in America—in which the contestants are given the *answers*, and then they need to figure out the *questions*. That's a lot tougher than you would think, as I've noticed from watching the show.

"So I wonder what you're cooking up, Hanna?" Harold added, with a chuckle.

"Try this answer—*flicka*," smiled Hanna.

"I think I'm trapped," admitted Harold. "But I'll say that the question is—What's the biggest challenge in my life now?"

"Good luck in sorting that out!" declared Karl, as they all, including Harold, laughed at Harold's romance challenges.

ARC of CONFLICT

Chapter 11: Class Act

On Wednesday morning, a Swedish breakfast of bran cereal with yogurt, plus boiled eggs, toast and vegetables stoked up the presenters for the classroom challenge.

In the classroom, the trio in front of the 25 upper-level students, quickly got their attention, especially with Harold towering over both Karl and Hanna.

Hanna explained to the class that she was capitalizing on an opportunity to hear from two visiting specialists in the petroleum industry. "Go easy on them, because they hadn't anticipated this opportunity to confront—or be confronted by students."

Several of the students smiled, perhaps in anticipation of a tangle with representatives of industry.

Hanna introduced Karl and Harold as managers in the petroleum industry, based at Stavanger. Mr. Onstad coordinates the complicated instrumentation system vital to Norway's Statoil, and Dr. Shostrom manages the output allocation for an international consortium of petroleum marketers.

Oil platform

"I think he hopes to be another Rockefeller," she chuckled.

"Both men," continued Hanna, "have faced the rigors and dangers of life on an oil platform in the middle of the turbulent North Sea. And previously Dr. Shostrom dealt with other

underwater challenges and threats, including service as a U.S. Navy Seal."

Karl led off the presentation with stories of the importance of precise control of the flow of oil, and he described bouncing around in the control room on the oil platform during a storm at sea.

"Like being on a ship, hoping that it doesn't sink. Which has happened to oil platforms, by the way," he declared. "Getting to the platform can also be a harrowing adventure, by ship or helicopter, halfway to Great Britain."

"When will you use your leverage to lower petroleum prices?" asked a man in the class.

"We see increased competition on the horizon," answered Karl, "and international output should lower prices. The Middle East watches the pulse of production carefully, South America offers increasing potential, and the United States has great output, albeit great appetite too."

Harold entered the interaction to remind the students of the taxes on top of taxes that make the cost of gasoline high. "In the United States, we have federal, state and local taxes that cost more than the product, and I presume you do as well."

"At least," said Karl as he got back in the discussion, "Norway is hanging on to much of the profit as a form of insurance for the time when the oil might run out, so our country in a way has kept prices artificially high for that reason."

Before Harold could describe his marketing responsibilities, an attractive coed asked Harold what he did as a Navy Seal. "Mighty dangerous, isn't it?"

"Maybe it's in my Swedish roots," smiled Harold. "After all, the Eastern Vikings were the Rus from Sweden.

A Rus (Eastern Viking from Sweden) at Constantinople

"As daring and determined merchants, they sailed and portaged their way as far as the Black Sea to do business.

"And those Rus became the Russians, so does that link your economy to the Soviets?" asked Harold, with a grin.

"Sometimes it seems that way," laughed a man in the front row.

"Maybe my heredity plus experience prepared me for the danger I've lived with, including eventually diving under an oil platform to survey the structure—to reassure my clients.

"I'm from the state of Washington, and I learned to go up as well as down, by climbing some huge trees there. You have a lot of forests in northern Sweden, but I doubt that those trees come close to the size of the timber in the American West. I also got a closer look and feel for at the size of the logs when I worked in a sawmill one summer.

"I learned a lot about life and work in the forest and mill, and I also learned what I didn't want to do the rest of my life.

"By the way," added Harold, "our Boeing Company from Washington State claims an unofficial stake in your SAS.

"Evidently, some of the B-17 bombers of World War II couldn't make it back after being sent on a run over Europe from Britain, so they landed in Sweden. And Sweden assessed a fine and confiscated the planes to make them part of your airline.

"Good catch," I'd say.

For more history of aircraft, including the Boeing B-17, Harold recommended a visit to the Museum of Flight in Seattle.

"Another job I didn't want to do the rest of my life involved hauling in salmon on a commercial fishing boat up near Alaska. Rough water…but good pay in the summers while I was a student at the University of Washington. That certainly motivated me to study…and get my bachelor's and advanced degrees. Books never looked so good!"

"Guess that prepared you for the wild ride out to the oil platform," a student called out. "Do you get seasick?"

"Sick of the sea, you might say," answered Harold, with a slight smile, "despite my Swedish name meaning 'sea' and 'stream'.

"The sea called me again when the U.S. military wanted me," added Harold. "That's when I became part of the Navy Seals. Most of the recruits for the Seals drop out the first week—to give you an idea of that challenge. I survived the training…and the dangerous duty too. But once again, books looked mighty good in comparison.

"Now, the North Sea often seems worse than the Pacific where I trained as a Seal at San Diego.

"Then," reflected Harold, "I think of the Norwegian explorer, Thor Heyerdahl, and my admiration for how he and his crew coped, adrift on their raft in the Pacific. At least, when I met him on his later visit to the Pacific Northwest, he could look back with amazing calm. A bit like I do when I think of my survival at sea. Secure in my survival."

"Doesn't the name 'Pacific' mean peaceful?" questioned a woman in the class.

"Some of the time, but it's so big that it includes lots of storms. Even a tsunami once in a while," answered Harold.

"I've heard about those—racing across the ocean and building massive power," said a student.

Thor
Heyerdahl

"Sometimes even reaching the West Coast of the United States," Harold indicated.

"But now I'm pretty much of a desk jockey in the Stavanger office," he added.

"Desk jockey…American slang?" asked a young man. "We're adding a load of slang to our language, mostly American. Makes it a lot tougher to talk English, though. Always trying to figure out the latest addition to the slang vocabulary."

"So am I," laughed Harold. "Now we have all the technology talk coming along too. And Hollywood adds and creates slang too."

"Here's a different kind of economics question," stated a man in the back. "I wonder, by your size, did you get your way paid through college as a so-called student-athlete?"

"Believe me," Harold responded, "I earned my scholarship… with the constant football practice, trying to study while traveling to games, worrying about the risk of injury, uncertainly about my future on the team, coping with constant criticism…from the fans, alumni, sports writers…and students like you.

"I guess I should consider it practical preparation for the uncertainties in the world of economics…and the economics of the world," added Harold.

"Now it's my turn now to pick on an American," said Hanna. "Don't your people get a guilty conscience from using such a disproportionate amount of the world's resources…including petroleum?"

"A twinge, perhaps," said Harold, "but our country has grown accustomed to taking what it wants…since the earliest of European immigrants…who came for religious freedom…but only for themselves. And the land seemed plentiful, so they took it."

"Is that how you justify your manifest destiny as you pushed the natives out to ensure easy taking?" Hanna asked.

"Just think back to your area—Scandinavia—about a century ago," countered Harold. "One-fourth of your population left because of starving conditions, so the land and resources in America must have looked like paradise. The natives there didn't seem to use the land well, so the immigrants did.

"Your writer Vilhelm Moberg depicts the desperation of farmers in Smaland, in contrast with the hope his immigrants found in America. Ole Rolvaag offers a similar story about Norwegian immigrants.

"With incentive and determination, most of them succeeded in their new land. And their inventiveness and industriousness enriched America."

"I, for one, commend you," said a young woman, with a hint of a smile, "at about the same time as I criticize you. But I know your country has been generous…such as the Marshall Plan for our rebuilding after Europe's self-destruction.

"And Scandinavia in particular has benefitted in so many ways and for so many years."

"One last question from me," Hanna added. "We wonder and worry about the negative impact of Hollywood. In fact, you look like you were in the movies yourself…like a double for John Wayne."

"Considering this is a course in economics," Harold answered, "I imagine you understand the financial impact of Hollywood.

"Huge business, especially in sunny southern California. Besides, show biz seems to be good business here in Sweden, including the many actors you lend to Hollywood.

"Maybe the most famous from Sweden was Greta Garbo, who fits the image of the Swedish desire for privacy," Harold speculated, "though the hope for privacy seems like a contradiction for an actress."

"Considering that my background is Norwegian and German," interjected Hanna, "I find it interesting that the famous Ingrid Bergman is Swedish and German."

"And I remember Swedish actress Signe Hasso," Harold mentioned, "because she portrayed German women, including a role depicting a Nazi spy."

"Then we have our Norwegian actress, Liv Ullman," declared Karl, "who was the Swedish wife of Swedish actor Max Von Sydow in the movie about *The Emigrants*. Great reminder of the challenge of change by Swedish author Vilhelm Moberg."

Greta Garbo   **Ingrid Bergman**   **Signe Hasso**

Norwegian actress Liv Ullman and Swedish actor Max Von Sydow portrayed Swedish emigrants in the story by Vilhelm Moberg.

"But I agree that Hollywood and TV have been corrosive in society," said Harold as he concurred with the concern expressed earlier, "so maybe a personal boycott could be your only vote against it. And you'd have my vote too."

"Before we finish," warned Hanna, "I did threaten Dr. Shostrom with a Jeopardy! challenge—you know, I give the answer, he states his question…as the students smiled in anticipation.

"So the answer I offer is *Statue of Liberty*," said Hanna.

"Easy," said Harold. "What important sight will you see when you enter the harbor of the New York City?"

"I take that as an invitation," chuckled Hanna.

"I'll be happy to be your guide," responded Harold, with a smile.

"Now, let's thank both of our guests for sharing," Hanna urged, and the students clapped vigorously.

Then, as other students crowded around Harold, the student from the back of the room approached: "Sorry for my sniping about your athletic scholarship. We hear many stories about the major investment in college sports in America.

"But you offered a good answer. And you provided intriguing and insightful information about changes in our economy.

"So, thanks.

"And I hope you enjoy more of Sweden."

"And I hope you enjoy America sometime," answered Harold.

Chapter 12: Recoup & Regroup

"You don't look too much the worse for wear," declared Greta, when the trio returned to the farm from the College.

"Oh, Harold may never want to speak to me again," announced Hanna, with a restrained smile. "I berated his country for greed and gluttony, you might say."

"At least I got off easy. Nepotism," Karl chuckled. "As your cousin, Hanna, I imagine you avoided a family disagreement by focusing on America's tendency that 'might makes right'.

"But as a Norwegian, I didn't want to remind the students that Sweden has had a history of privilege for the powerful. The peasants of the not-so-distance past suffered from the whims of the royalty and the aristocracy. And the peasants faced conscription for the military adventures of the leaders too."

"Touché!" said Hanna. "And my apology, Harold, for my needling."

"I enjoyed the exchange…rare privilege with such a dynamic and attractive professor," smiled Harold. "Your questions provided me an opportunity to share my thoughts, pro and con, about America. I hope I didn't ramble on too long, though."

"Not at all," Hanna responded. "The students seemed attentive and eager to learn from another angle of economics…and so did I. And they really perked up with your mention of Hollywood…plus your doppelganger link to John Wayne."

"Well, Cowboy," laughed Greta, "I look forward to your skilled assistance when it's time for me to round up my cattle, wandering around on my ranch. Are you good with the lasso and branding iron too?"

"Maybe we could at least get you an Australian sheep dog for your roundup," responded Harold.

The thought of an Australian sheep dog in Sweden met with approval, as the Onstads agreed to the need—if they got a herd of cattle to go with it.

But Hanna hinted to Harold that a real American cowboy would serve the Onstad "ranch" even better. And make a good companion besides!

"Yes, a sheep dog would make a good companion," Karl added, as Hanna nodded in agreement. "As would a cowboy," he chuckled.

"Just think," laughed Hanna, "our own American cowboy! Can we sign you on, John Wayne?"

"On the other hand," said Greta, "I admit that an Australian sheep dog would be a bit of a novelty for me and my neighbors.

"But so would an American cowboy," she admitted, with a wry smile.

"Maybe if we strike gold by following your treasure map," said Karl, "you could afford one…a sheep dog that is, with a pedigree of course!"

"Well, sorry, but we do need to get back to the reality of that mysterious treasure," interjected Hanna. "What steps do you see next, Karl?"

"I suppose first we'll wait to learn what, if anything, Kari has managed to find at her printing company," he stated. "If we hit pay dirt there, we can compare that material and information with your copy of the 'will.' Still just a bunch of 'if's' for now.

"Your notice, Harold, of the arcs on the poster at least gives us a glimmer of hope that leads to the bunker on Karmøy. Then more maybe's."

"Well, maybe for now we can gather for supper…around the campfire and next to the chuck-wagon, of course," laughed Greta. "At least we will share some pot-roast, beans, veggies, spuds…and homemade bread."

"Wow! Homemade bread!" exclaimed Hanna. "You do have some control of your time, now that you're at least semi-retired."

"Strawberries and cream for dessert," Greta added. "But no, I didn't milk the cow and separate the cream. Though I'll have you know that the cream separator was invented here in Sweden. So I may give it a try someday. After I get my trained cow-dog," she laughed. And so did the others.

"Now that you mention the cream separator," Hanna declared, "I should remind you, Harold, in case you don't know it, that Sweden's Alfred Nobel made a big impact—if you'll pardon my explosive expression—in the United States…and around the

World with his international prizes for the sciences, medicine, literature, peace…and appropriate to me…economics. "

"He was a contradiction from someone who really wanted to be a poet, not a manufacturer of dynamite!" emphasized Hanna.

"Oh, yes," exclaimed Harold, "we Americans are well aware of the Nobel honors. You may know that our chemist Glenn Seaborg, who is also of Swedish descent, by the way, helped in the development of the Atomic bomb that forced Japan to surrender to end the war in 1945."

Nobel-prize-winner Glenn Seaborg at the University of California, Berkeley.

"We Americans thank another Swedish scientist, John Ericsson, who invented many components of the revolutionary ship

Monitor, which helped the Union preserve our country during our Civil War," explained Harold.

"And," added Karl, "I understand you can connect here with some of his work—because his father was an engineer involved in designing the Gota Canal that crosses Sweden.

"Maybe we can link up with that sometime, to experience the Canal, and pay tribute to the Ericsson inventions."

John Ericsson's *Monitor* was designed to be mostly submerged, with only the top part visible.

"I'm for that," exclaimed Hanna. "A good reason to enjoy a cross-country trip to savor Sweden."

"That could also be a good excuse for us to go to New York," suggested Harold, "to see his statue in Brooklyn...besides seeing the Statue of Liberty in the New York harbor."

"Well, listen to those plans," smiled Greta. "Makes me envious...but sounds like signs of a budding friendship."

"Hmmm," smiled Karl. "Maybe we'll get an American connection in our changing family. So I vote yes for that," he laughed.

"Hold your horses," Hanna declared. "Let's talk more about major change. Imagine running the gauntlet in my economics class by trying to explain radical societal changes in recent decades in Sweden. For example, not too long ago, the country was divided between aristocrats and peasants. Guess what that eventually caused?"

"Emigration, like in Norway," said Karl.

"I presume the affluent Swedish upper-class finally got the message: change or wither," explained Hanna. "So gradually the country changed, with expanded education, incentives for business, supportive social services, and wider political involvement.

"Just think, for example, of the amazing influence of two recently prominent aristocrats.

"I believe you could say that the most dramatic was Raoul Wallenberg. Though he was part of the richest banking family in Sweden, he chose to risk his future…and his life…in helping thousands of Jews in Budapest evade the grasp of the Gestapo near the end of the war. Then he mysteriously disappeared…presumably at the hands of the Soviets.

"Another, who lost his life mysteriously as an advocate of peace, was Dag Hammarskjöld. He was the second Secretary-General of the United Nations, succeeding Trygve Lie of Norway.

Raoul Wallenberg   **Dag Hammarskjöld**

With those examples, they all agreed that positive change can happen, on a small scale in families and communities, as well as vast changes globally.

"Maybe we can sleep on that agreement," smiled Greta. "You might welcome rest after running the gauntlet at Hanna's class.

"Well, I certainly have enjoyed your visit," said Greta. "I hope you will come back!"

"Now, in penance for attacking America," said Hanna, "I will offer you another massage, Harold, with the hope you will sleep well."

"Oh yes, that feels good," murmured Harold as the massage took effect.

"Maybe," said Karl, as his last word of the day, "we'll come back with your mysterious treasure…or at least with more information about the mystery."

"And I hope I can return for another massage," Harold said quietly, "not to mention learning more from your grasp of economics, Hanna…and your understanding of life's challenges."

So the next morning, Karl and Harold again boarded the Statoil plane with the homeward bound executive.

He was surprised to learn that they had appeared as presenters at a Kristianstad College economics class.

"Hope you put Statoil in a good light," he hinted, "because an economics class could hardly avoid the topic of petroleum."

"As an American," explained Harold, "I took the most heat from the professor and some of the students. They expressed a legitimate resentment about our high consumption of resources.

"But I did counter by declaring that we do also serve the world in a far greater way than any other country."

"Maybe," said the executive, "we'll have to share your thoughts and those from the College classroom with our leaders at Statoil. We do feel the heat of negative public opinion too."

"The tone and the range of the reactions in the classroom might be revealing and worth hearing," said Harold. "Let us know," he added, as he provided his business card.

As the plane moved closer to home, Karl and Harold focused again on the mystery they would soon be immersed in.

"At this stage," said Karl, "much will depend on Kari's findings, if she learns anything from her company's storage of wartime documents."

As they touched down, Kari waited in the Volkswagen at the airport. She hopped out and dashed to meet them when they walked her way.

Her big smile signaled more than a welcome home, as she quickly announced some dramatic news.

After a hug and kiss for Karl, and a hug for Harold, she summarized the results of her research: "Yes, like you two said, the Germans did keep careful records and material. We didn't find a printing plate like you described, Harold, but we did find the papier mache mat that could have been used to make a plate.

"Strange, though, it seems too clean to have been used for casting lead. I'll leave that up to you to determine, Harold, if it even has a bearing on the mystery.

"But I'll try to find a record of heavy printing plates being brought by Col. Schorn to Stavanger. Meanwhile, we have saved several copies of his posters that we found," said Kari.

"I suppose," speculated Karl, "as you said, Harold, the printing plates probably would have been melted down to use the lead again, as long as that mat was available for another casting."

"Seems logical," admitted Harold, "but still we have the puzzle of the arcs scribed on the poster used as the improvised will.

"Do you have one of the posters from that file, Kari?" asked Harold.

"No, but we can get one from the file we started at our company," she answered. "And the papier mache mat too."

"Good!" continued Harold. "We can check to see if the strange arcs we saw in the poster Greta had are also on the posters here. If not, the mystery continues."

"What arcs are you talking about?" Kari asked.

"Sorry," said Harold. "I forgot that you haven't seen the poster that was modified to be a document like a will. Near the signature by Col. Schorn, we saw two arcs under his name. We wonder if the arcs might be a clue, indicating a connection with the gun emplacement we saw on Karmøy."

At lunchtime the next day, Kari pulled a poster out of the file to show Karl and Harold. And she got the papier mache mat out too.

"No arc inscribed on the printed poster," said Karl, "and not other writings or signatures either, of course."

"This mat for casting does match the proof, so it could have produced a lead plate for printing," said Harold.

"Guess that gives more credence to the arcs alongside Col. Schorn's signature, and he certainly would have known of the arcs created by the gun emplacement," Harold surmised.

"Yeah, after all," Karl emphasized, "he no doubt designed that bunker and would have known that outline of his own design."

"Kari, does your file indicate when the concrete was poured for the bunker?" asked Harold.

"Here's something," she said, "in German, but I think we can figure it out. The record indicates the concrete was poured about three months before Col. Schorn left for the coast of the English Channel, according to what I learned about him earlier."

"So he could have buried something in the wet concrete," concluded Karl. "And he was never able to come back for it."

Chapter 13: Haystack Search

"Yup, I agree with you, Karl, this will be like looking for a needle in a haystack," Harold muttered. "And we don't even know if it's a needle."

"The good news," reported Karl, "is that Kari learned from Norwegian officials that we have permission to search the bunker. If we have reasonable indication of something hidden, we can pry out electrical junction boxes, for instance. At least they could be put back."

So, to avoid attention from anyone who got wind of the search, the foursome of Karl, Kari, Harold and Sofie went on another picnic on the bunker.

While Kari and Sofie put on the pretense of the picnic, Karl and Harold initially pried out junction boxes…without finding any sign of treasure of any kind.

Then they checked the floor…no success. Then the gun port, to no avail.

Karl and Harold admitted to Kari and Sofie that the haystack revealed nothing so far.

"For Greta, we'll keep on looking," said a determined Karl.

"I'm with you," confirmed Harold, and Kari and Sofie agreed to provide continuing cover of the activity with some form of picnic.

"At least I like your food," laughed Harold. "And I enjoy being with all of you on this venture…crazy, but maybe not hopeless."

"I imagine folks who see us coming here just think we're lovebirds, looking for privacy," said Karl. "And I guess they're right," he laughed.

"Fortunately, that makes a scene that's easy to believe," said Kari. "But with Sofie, we have to be careful so her boyfriend from Trondheim doesn't show up at the wrong time."

For the next Saturday picnic, they came armed with four small ball pein hammers. As soon as they completed the pretense of a picnic, they began tapping the walls and ceiling of the main dome for any strange sounds, like an unexpected cavity in the concrete.

"Any sandwiches left?" asked a discouraged Karl, as he sat on the floor.

"I believe so," said Kari. "I've brought some for all of us. We need some nourishment…and encouragement…and water so we don't get dehydrated."

"Hold on!" shouted Harold just then. "This doesn't sound right," he said quietly. "Listen," he said as he tapped in an area of several square feet. "Not hollow, just a dull thud that's not like the rest of the concrete. Probably material that's a lot denser."

"Maybe some metal reinforcement," suggested Karl.

"Doesn't seem needed right there," replied Harold.

"Hardly the needle," commented Kari.

"Maybe the mysterious treasure is large, and the concrete wall would make a good hiding place," Harold said as he thought aloud. "After all, Col. Schorn supervised the construction, and he could even have buried something big as the concrete was poured. Or several large items."

"Now," Karl declared, "I think we should get out of here before our noise attracts attention."

"Yeah," Sofie chuckled nervously, "it would be a bit difficult to explain that we always pound on walls when we have a picnic."

"I've got an idea!" exclaimed Harold. "And I could kick myself for not thinking of this before," Harold then muttered, as the others waited for him to explain.

"Let's get out of here and gather again in Stavanger," Harold announced. "I have a theory related to my printing experience, but right now I worry that someone else might get curious...and get lucky and detect that strange sound in the wall."

"My place would be good," Kari offered. "I'll make some coffee and I'm sure I have some dessert."

In the Volkswagen, they talked little and seemed skittish, as if someone could read their minds and find the treasure... whatever it might be. And three of them were eager to hear the rest of Harold's premise.

The sound and smell of the bubbling coffee calmed them, and so did the chocolate cake.

"German chocolate cake again," laughed a tense Karl.

Finally, as they gathered around the dining table, Karl asked—almost ordered—Harold to "spit it out!"

"Okay," Harold uttered with a higher voice than usual, "I think we may have struck gold!"

"Quit your kidding!" shouted Karl. "At a time like this!"

"Not kidding," Harold answered, as the others waited impatiently.

"Remember, Karl, we learned that Col. Schorn's family had a large investment in gold," explained Harold. "If he wanted to take that out of Germany, just how could he do it?"

Molten metal being poured into a casting case: "For Col. Schorn, did the metal being poured turn out to be *gold* rather than the lead typically used for printing plates?" wondered Harold.

Harold answered his own question: "I think he might have turned that gold into printing plates. Not lead as we talked about before. His family had a printing business. So he could have melted the gold and poured that gold into a casting case like we used at the Seattle printing company. So, instead of bringing plates made of lead on the pretense of printing more of that poster, he brought printing plates cast with gold! Probably darkened to look like lead. Several of them, worth a small fortune! Willed to Greta! *On a poster*!"

"Ingenious," said Karl, stunned and quiet. "Seems logical, and I sure hope you're right. Now how do we go panning for gold?" pondered Karl, in a happier mood.

"I don't think we should continue tapping or trying to cut into the concrete until we have a better picture of the plates," declared Harold.

"I've got it!" Karl, with renewed excitement, blurted out. "Among all the instruments we use for Statoil, we have portable scanners that can search, like an x-ray, even through concrete. I think we could possibly get a reading about how deep the plates might be in the concrete."

"Hold everything!" interjected Kari. "While you guys get ready to carve concrete, I think we should check out Norwegian laws related to salvaging this possibly tremendous find…and the rules about inheritance. As I recall from other situations in our company, we've dealt with the Norwegian Office of Inheritance. I could start checking. But we don't want to let the cat out of the bag until we have our ducks in a row," she added, as they all laughed at the mixed metaphors of mixed species.

"Don't forget," Sofie reminded the others, "Germany and now Sweden might want involvement…at least regarding tax questions."

"Then there's the likely emotional impact on Greta and Hanna," Karl said. "Imagine the potential publicity about this sensational find, combined with dredging up Greta's past during the German conquest and occupation of Norway."

"We also need to seek justice in this issue by involving the National Monuments Commission…for permission, legality, maybe even for help in extracting the plates," added Karl.

"If there really are the gold plates we visualize," Harold cautioned.

"I think, for now, we should put this on hold," stated Karl, "until we resolve those issues. Sorry to suggest a delay, just at the peak of our excitement about this discovery."

"Makes sense," Kari agreed reluctantly, as did the others.

"Meanwhile, we can divide the territory," suggested Karl. "Kari with the Norwegian inheritance questions, Sofie the Monuments Commission's role, Harold the Swedish regulations with input

from your relatives, and me for interaction with Greta and Hanna about their concerns and wishes.

"Meanwhile, Harold and I will discreetly scope the bunker with the scanner that I can borrow."

"Maybe," suggested Harold, "that will confirm our assumption that—as prospectors must have said in Alaska—there's gold in them there hills."

"Oh no!" laughed Sofie. "First we hear the cowboy talk from you, Harold. And now we get the sayings from Alaskan prospectors. Right out of Jack London, I suppose."

With a repressed grin, Harold responded: "Well, to be serious, if there is gold," said Harold, "we will definitely need help from the police and even a military unit to guard our 'buried treasure,' which I guess might be worth up to a million in U.S. dollars, and bigger numbers in Scandinavian money.

"Meanwhile," he added, "you may have heard the World War II warning: *Loose lips sink ships.* So we'd better keep this from the news media…and everywhere else that has ears."

Chapter 14: Tying Loose Ends

Trying not to be conspicuous, Karl and Harold returned to the bunker and scanned the wall that sounded like it had dense material embedded inside—and did indeed confirm their suspicions about "buried plates."

Then the team members proceeded on their individual assignments.

Most of the research could be done right in Stavanger, but Karl and Harold saw the need to travel to Kristianstad again for a conference with Greta and Hanna… to report their good fortune, but also to weigh the consequences of the publicity it would generate.

They felt Harold's relatives, the Olssons, would be desirable additions to the discussion later, after Greta and Hanna worked through the initial shock. They assumed that as business persons, the Olssons probably could provide general advice about this kind of inheritance.

Also, the Olssons had hit it off well with Greta before, so they might offer a practical and emotional sounding board.

"Ironically," reflected Karl, with a wry smile, "driving would take a lot of gas… and time… going south around Norway, across on the ferry, then more driving across Sweden."

"Cheaper if we went on my motorcycle," chuckled Harold. "Usually good weather this time of year."

"No thanks," Karl laughed, "so I guess, as you cowboys might say, we'll have to 'bite the bullet' and go on a commercial flight. No hops available that I know of."

So, again, Greta and Hanna met them at the Kristianstad airport.

"I thought my economics class and I might have scared you off," said Hanna. "Especially with our attack on America."

"Oh, your massage more than made up for the attack," laughed Harold. "Had to come back with my raincheck from my masseuse. Now I need tender loving care because our small plane pinched my shoulders."

"Okay, I'll happily honor that raincheck," smiled Hanna.

"Great to have the two of you back," declared Greta. "And, Karl, you hinted on the phone that you have exciting news. Must be significant for you to fly here to share it. About the poster, I imagine."

"We think the treasure implied by the poster might be real," Karl responded, with excitement rising in his voice. "Remember the arcs marked on your poster? We concluded that they indicate the curves of the gun emplacement on Karmøy.

We speculate that Col. Schorn's involvement in the design and construction would have allowed him to bury his family's gold in that concrete.

As a shocked Greta remained silent, Hanna began the questioning: "How could he have possibly brought his family's gold out of Germany and into Norway, under the noses of the Gestapo?"

"Remember," Karl reminded Hanna, "that Greta talked about Col. Schorn's comments about the heavy printing plates for the poster? They didn't show up in storage at the printing company, so we speculated that the printer had melted them down to use the lead again."

Karl continued his report: "But now, we're quite sure we've located the plates—buried in the concrete of the bunker on Karmøy!"

"Why on earth would he bury printing plates?" asked Greta, recovering from her earlier shock.

"Because," explained Karl, as he paused for dramatic effect, "*they might be gold, not lead*. Probably darkened to look like lead to make them appear to be ordinary lead printing plates when he brought them to Norway."

"Just think of the irony of this situation as it relates to Johannes Gutenberg. He invented movable type made of lead—while he worked for his rich relatives in Germany who were goldsmiths."

"So *we might inherit that treasure, after all*," declared Greta, with a large grin. "A wry twist for Hitler and his Gestapo."

"By now those plates must be worth a fortune!" exclaimed Hanna.

"In America," Harold explained, "the printing plates—probably eight of them—might be worth about a million dollars. I'll let you translate that into Kronor."

That announcement stunned Greta again.

"Oh!" she exclaimed, "I need a cup of coffee…and lefse."

"I'll make it," said Hanna. "Harold, maybe you should help me. My hands won't stop shaking."

"Glad to oblige," Harold responded and he held Hanna's hand while they walked together to the kitchen.

"Umm, that felt reassuring," said Hanna, with a smile. "Worth repeating, like my massage. Comforting."

Then, in the other room, Karl announced to Greta: "We thought you might want to get some local financial and emotional support. And we were impressed by the Olssons. Maybe they would be helpful, if willing."

"Since we met them when you were here, we've become good friends, partly over our farming challenges and sharing knowledge about making the best of retirement.

"I'd vote in a minute about seeking their support," she added.

"If all proceeds as we anticipate," explained Karl, "we do think you could handle the money. But we worry about how the obvious publicity might affect you and Hanna."

"Yes," said Greta, "my affair with a German soldier would make headlines. Embarrassing to me, Hanna, and our family in Norway. But maybe the true character of Ludwig would surface. Here comes Hanna with our *fika*," said a relieved Greta. "Let's review that publicity quandary with her, and give her a chance to react personally."

So Karl repeated what he had warned Greta about…the likelihood of extensive and embarrassing publicity, once the story of the gold gets out.

After setting the coffee and lefse down, Hanna grabbed Harold's hand for security as she sat down.

"I'm still in shock about the treasure," she admitted, "but this new slant forces me to think further about our life here. And about my reputation as a professor. About my relationship with my students and everyone else at the College."

"Karl suggested that the Olssons might provide helpful counsel about the financial and the personal impact of the situation," said Greta.

"Good thought," confirmed Hanna. "We've come to adore and appreciate your relatives, Harold. They are so resourceful, as well as kind and helpful. And I'm sure they would help us through our state of shock too."

"I'll call and surprise them again," smiled Harold, "and invite them over as soon as they're available, for a special conference."

"Please do," said Greta.

"What a conference this will be," Hanna smiled. "Love your relatives," she laughed, "and I'm getting fond of you too."

After he made his phone call, Harold reported, "Emil and Edith look forward to seeing Karl and me again, and joining us for *fika* again.

"I did warn them that this would be a business conference as well as a social gathering with the four of us. I hinted that I hope they can be of help with a significant development for both of you, Greta and Hanna.

"I suspect they can hardly wait for morning, after I threw in that bit of mystery," laughed Harold.

"Cruel," laughed Hanna. "Now they won't be able to sleep tonight…and I don't know if we will be able to either. No class tomorrow, so I think I will just stay here tonight and enjoy the solace of the farm. No city noise, just the breeze…and sounds of the animals."

"Haven't experienced that for a long time," smiled Harold.

"Come out on the porch and sit a while, as you cowboys say in your Westerns," chuckled Hanna.

"Enjoy it," said Karl, "while Greta and I review some steps to consider for patching up family relationships. I know that Jakob, for one, will be overjoyed—and he'll be shouting 'I told you so!' as he brags about his belief that a hidden treasure existed. There'll be no living with him now!"

As Hanna sat in a rocking chair on the porch, Harold volunteered: "My turn to offer you a shoulder massage to ease your tension."

"Wonderful offer," said Hanna, as she leaned back against him. "Thanks for everything, including helping me handle the initial shock."

"You are a sterling person," said Harold, as he leaned down to kiss her neck.

"Gold, not sterling," she chuckled. And she turned her head to kiss him on the cheek. Then he put his hand on her chin and kissed her on her lips. So she reached back to pull his face tight against hers.

"I think I'll be able to sleep well tonight," she whispered, "with fond thoughts about you—the cool American with the warm kiss."

Chapter 15: Welcome Advice

The next morning, the Olssons arrived early for the *fika*.

"Thirsty," laughed Edith. "And eager to see cousin Harold again."

"Sorry," said Harold, "to let you wonder and worry, but I'm sure you'll appreciate the mystery when you learn of a possible financial development for Greta and Hanna, even though it comes with some personal challenges too."

"Let me summarize," said Karl, "and you three can chip in with any details I miss, Harold."

So, as the Olssons sat in amazement, Karl recited Greta's story from the war and about a new chapter evolving now in Norway.

"Edith and Emil," said an emotional Greta, "when I recover from this amazing turn of events, we should talk about our dream of starting a gallery to display and sell art, including your glass designs and screen prints."

"Great to hear that positive intent," said Harold, "but for our immediate concern, we think Greta and Hanna will need advice and emotional support at this stressful time. I offered you, my helpful relatives, to be neighborly counselors."

"Thanks for that, Harold," said a choked-up Edith. "We've become good friends, and I think we have knowledge that could be helpful."

"For one thing," said Emil, "from my experience, I think such a special inheritance will not cause any ownership or tax consequences here in Sweden."

"Thank you for that kind of reassurance," expressed Karl. "We hoped we'd find that kind of positive support here."

"We're worried about how people here in Sweden will react to the past scandal," Hanna said, with an emotional quiver.

"Like Swedes," chuckled Emil, "with quiet acceptance and understanding."

"I hoped and expected that," said Harold. "Thanks for confirming it.

"Now I visualize the power that Greta will command from the gold plates. Reminds me of Mark Twain's story, *The Million Pound Bank Note*. For a short time, the temporary holder of that note didn't need to pay for purchases—because all were glad to extend credit to him. It was as good as gold."

Then the ring of the phone interrupted, and Hanna answered. She told Karl that the call was for him from Kari.

Then the others heard him: "Hold on. Slow down, so I can take this in and share it.

"Wait a minute," he said to Kari, as he asked: "Does this have a speaker-phone choice so all of you can hear?"

"Press this button," said Hanna, as she pressed the button for Karl.

"Okay," Karl said to Kari, "all in the room, including Harold's Swedish relatives, can hear you now. So boil down your report as best you can, so you don't have to run up a big phone bill."

"Don't worry," replied Kari, "I'm in the conference room at the office with several from the company, and we are on a speaker phone too.

"First, the amazing and reassuring news about the hidden printing plates, Harold. Though the plates haven't been extracted from the concrete, the team of specialists involved did drill through the concrete. And they struck gold, just as you predicted."

"Hurray! Hurray!" shouted Karl, echoed by all the others.

"Our company officials are here, because the company has taken charge of all the security, legal issues and publicity aspects of this amazing event," announced Kari on the phone.

"They feel concern for you, Greta, and Hanna. So, regarding publicity, they think they can turn that into a positive story, particularly about how Col. Schorn even fooled the Gestapo."

"And kept the family gold from Hitler," chuckled Karl.

Then Kari added: "The officials here would like to have photos available, with the hope that you, Greta, might have a portrait or even a snapshot of Col. Schorn.

"The publicity staff would like photos of you and Hanna, and they will arrange to take photos here, as well as movies, of you and Hanna when you come back to Norway to get your inheritance."

"Yes, I do have photos of Ludwig," said Greta. "Even a photo of the two of us together."

"The company leaders think sympathy will turn your way, Greta, when the public learns about Col. Schorn's kindness to you and your students. And information will be revealed, Hanna, that before you were born, Hitler ordered your father to go down by the English Channel to build more defenses in anticipation of the Allied invasion. And there he lost his life."

119

"Now, Karl, and you, Harold, are being asked to return as soon as you can for the ceremonial extraction of the printing plates. Probably the most expensive printing plates in history," chuckled Kari. "The two of you will probably become notorious…well, famous.

"Gradually, Harold, even folks in Seattle will learn about how your work at a printing company there helped solve this golden mystery.

"And I need to alert you to another change, Harold. Brace yourself—Sofie and her boyfriend from Trondheim—I'm sorry to say—are back together. But they're doing their part in preparing for the dramatic return of the Norwegian heirs.

"And Greta and Hanna, you will be needed here also, of course. We know this will be a trying and emotional time. But, at least your return to Norway should be a positive reunion.

"First, you'll receive a warm welcome back," said Kari. "But also we want you, Greta, to bring the original document—your unusual will—so we can set the stage for you to receive your amazing plates of gold."

"I'll do that," responded a buoyant Greta. "And I'll bring Hanna along," Greta smiled, "and I'm sure she will be happy to console and comfort Harold for being jilted—if you want to call it that."

"Good personal support," chuckled Kari. "And, by the way," she added before she signed off, "don't worry, Karl…and Harold…about the trip back. Statoil will send a plane for you. As well as for Greta…and Hanna!"

The end of the beginning

Cast of Characters

Harold Shostrom of Seattle works in the petroleum industry in Stavanger, Norway.

Karl Onstad of Stavanger, colleague of Harold, works for the Norwegian organization Statoil.

Haaken Onstad is the father of Karl and Jakob and brother of Greta Onstad and uncle of Hanna Onstad.

Olga Onstad, husband of Haaken, is the mother of Karl and Jakob.

Greta Onstad, sister of Haaken, was the *de facto* wife of German officer, Col. Ludwig Schorn. Now she is a retired teacher in Sweden.

Hanna Onstad, daughter of Greta and Ludwig, is a professor of economics at Kristianstad College in Sweden.

Kari Larsson, girlfriend of Karl, is an employee of a law-book publishing company in Stavanger.

Sofie Stensgard, teaches school attended by Jakob Onstad, and she is a friend of Kari Larsson.

Marit Dahl, Norwegian assistant in Harold Shostrom's office.

Emil and Edith Olsson, Swedish relatives of Harold Shostrom, are neighbors of Greta and Hanna Onstad.

Afterword

Because I write extensively about Scandinavia-Americans, I'm always on the alert for a new and interesting slant to create another novel or story.

This novella evolved from the family history of my Norwegian-American wife, Anita **Hillesland** Londgren. The last time we toured Norway, we visited the site of the German bunker on an island where some of her relatives live.

Since then, we created most of this story based on other knowledge we've gained as a result of our continuing involvement in various Scandinavian activities. And, though we hope the characters and the incidents in *ARC of CONFLICT* **seem** real—they are just a figment of my imagination.

In this story, as in my other novels, I turn to my own experience and to my own advice about planning, writing and design as defined in my book *Communication by Objectives*.

My wife Anita (left, me on right) believes she is distantly related to St. Birgitta of Sweden, so our exploration of Scandinavia included the chance to see this statue of St. Birgitta at her abbey in Vadstena, Sweden.